LABORATORY EXERCISES IN PHYSICAL GEOLOGY

Laboratory Exercises in Physical Geology

Allan Ludman
Queens College of The City University of New York

Wm. C. Brown Publishers
Dubuque, Iowa • Melbourn, Australia • Oxford, England

Book Team

Editor *Craig S. Marty*
Developmental Editor *Lynne M. Meyers*
Production Editor *Marlys Nekola*
Designer *Christopher E. Reese*
Art Editor *Mary E. Swift*
Photo Editor *Diane Saeugling*
Permissions Editor *Mavis M. Oeth*
Art Processor *Amy L. Saffran*
Visuals/Design Consultant *Donna Slade*

Wm. C. Brown Publishers
A Division of Wm. C. Brown Communications, Inc.

Vice President and General Manager *Beverly Kolz*
National Sales Manager *Vincent R. Di Blasi*
Assistant Vice President, Editor-in-Chief *Edward G. Jaffe*
Director of Marketing *John W. Calhoun*
Advertising Manager *Amy Schmitz*
Director of Production *Colleen A. Yonda*
Manager of Visuals and Design *Faye M. Schilling*
Design Manager *Jac Tilton*
Art Manager *Janice Roerig*
Publishing Services Manager *Karen J. Slaght*
Permissions/Records Manager *Connie Allendorf*

Wm. C. Brown Communications, Inc.

Chairman Emeritus *Wm. C. Brown*
Chairman and Chief Executive Officer *Mark C. Falb*
President and Chief Operating Officer *G. Franklin Lewis*
Corporate Vice President, President of WCB Manufacturing *Roger Meyer*

Cover photo © Michael Lustbader/Photo Researchers, Inc.

Copyedited by Rick Nelson

The credits section for this book begins on page 228 and is considered an extension of the copyright page.

Library of Congress Catalog Card Number: 92-71003

ISBN 0-697-14706-1

Printed in the United States of America by Wm. C. Brown Communications, Inc., 2460 Kerper Boulevard, Dubuque, IA 52001

10 9 8 7 6 5 4 3 2 1

Contents

preface *vii*

Section 1
Minerals and Rocks

Chapter 1
Minerals 2

Exercise 1.1
Classification of Earth Materials 4

Exercise 1.2
Physical Properties of Minerals 4

Introduction to the Physical Properties of Minerals 5

Exercise 1.3
Identification of Minerals 12

Chapter 2
Minerals in Rocks 20

Exercise 2.1
Identification of Minerals in a Rock 22

Exercise 2.2
A Rock is More than the Sum of its Minerals 22

Exercise 2.3
The Origin of Typical Rock Textures 25

Exercise 2.4
Identification of Rocks as Igneous, Sedimentary, or Metamorphic 26

Chapter 3
Interpreting Earth History from Igneous Rocks 28

Exercise 3.1
The Importance of Texture in Studying Igneous Rocks 30

Origin of Igneous Textures 30

Exercise 3.2
Interpreting Igneous Textures 30

Identification of Igneous Rocks 33

Exercise 3.3
Identification of Igneous Rocks and Interpretation of their Cooling Histories 34

Chapter 4
Interpreting Earth History from Sedimentary Rocks 36

Formation of Sedimentary Rocks 38

Textures of Sedimentary Rocks 38

Relations between Grains 38

Exercise 4.1
Textures of Chemical Sedimentary Rocks 38

Grain Size 40

Grain Shape 40

Exercise 4.2
Interpretation of Transport Mechanism Based on Textural Features 41

Mineralogy of Sedimentary Rocks 41

Sedimentary Structures 43

Exercise 4.3
Graded Beds 43

Identification and Interpretation of Sedimentary Rocks 46

Chapter 5
Interpreting Earth History from Metamorphic Rocks 48

The Role of Composition 50

Exercise 5.1
Mineralogy of Metamorphic Rocks 50

Role of the Type of Metamorphism 51

Exercise 5.2
Metamorphic Textures 51

Role of Metamorphic Intensity 54

Classification of Metamorphic Rocks 55

Exercise 5.3
Identification of Metamorphic Rocks and Interpretation of Metamorphic History 55

Section 2
Earth's History Determined from Its Surface Features

Chapter 6
Introduction to Maps 60

Map Elements 62

Location 62

Exercise 6.1
Use of Latitude and Longitude for Location 63

Exercise 6.2
A Little Practice with the Public Land Survey System 65

Direction 65

Exercise 6.3
How Not to Get Lost 67

Distance 67

Exercise 6.4
Using Scale to Measure Distance on a Map 69

EXERCISE 6.5
Putting the Map Elements Together and Using Them 70

Topographic Maps 70

EXERCISE 6.6
Establishing the "Rules" of Contour Lines 73

Topographic Profiles 76

EXERCISE 6.7
Making a Topographic Map 78

EXERCISE 6.8
What to Do When Meeting a Topographic Map 79

Chapter 7
Streams and Stream-Carved Landscapes 81

Principles of Stream Erosion and Deposition 83

EXERCISE 7.1
Types of Streams 83

Stream Erosion: Downward or Sideways? 84

EXERCISE 7.2
Using Topographic Map Skills to Analyze Stream Behavior 85

Features Found within Stream Valleys 89

EXERCISE 7.3
Erosion and Deposition in a River Valley 89

Evolution of a Landscape Dissected by Streams 91

EXERCISE 7.4
Evolution of a Stream-Dissected Landscape 91

Chapter 8
Glaciers and Glacially-Formed Landscapes 96

EXERCISE 8.1
Comparison of Glaciated and Stream-Dissected Areas 98

Principles of Glacial Erosion and Deposition 98

Types of Glaciers 99

EXERCISE 8.2
Alpine Glaciation 99

Continental Glaciation 102

EXERCISE 8.3
Erosion by Continental Ice Sheets 102

Landforms Produced by Glacial Deposition 104

EXERCISE 8.4
Recognizing Landscapes Formed by Glacial Deposition 108

Chapter 9
Groundwater as a Landscape-Former and a Resource 114

Landscapes Carved by Groundwater 116

EXERCISE 9.1
Features of Karst Topography 117

Groundwater Resources, Pollution, and other Problems 121

EXERCISE 9.2
Farmer Jones vs. Grandma's Soup Company 122

EXERCISE 9.3
Groundwater Contamination 122

Chapter 10
Arid Landscapes and the Work of the Wind 126

EXERCISE 10.1
The Uniqueness of Arid Landscapes 128

Evolution of Landscapes in Arid Regions 129

EXERCISE 10.2
Examination of Arid Landscapes 131

The Work of the Wind 134

EXERCISE 10.3
Landforms Produced by the Wind 134

Chapter 11
Shorelines: Coastal Erosion and Deposition 140

EXERCISE 11.1
Factors in Shoreline Development 142

Waves 142

Coastal Materials 142

Progressive Erosion Along Shorelines 143

Emergence vs. Submergence 145

EXERCISE 11.2
Analysis of Shoreline Features 146

EXERCISE 11.3
The Greenhouse Effect and Its Impact on Shorelines 155

SECTION 3
INTERNAL EARTH PROCESSES REVEALED BY SURFACE FEATURES

Chapter 12
Interpreting Geologic Structures 160

The Relation between Topography and Deformational History 162

EXERCISE 12.1
Topographic Clues to Deformation 162

EXERCISE 12.2
Making a Geologic Map from Topographic Clues 165

Folds and Faults 168

EXERCISE 12.3
Analyzing Geographic Maps 169

EXERCISE 12.3
Completing Block Diagrams 172

Chapter 13
Interpreting Geologic History 182

EXERCISE 13.1
Criteria for Relative Age Dating 184

EXERCISE 13.2
Interpreting the History of Parts of North America 188

EXERCISE 13.3
A Pair of Challenges to your Abilities as Geological Detectives 199

Chapter 14
Plate Tectonics 204

EXERCISE 14.1
Fitting the Pieces Together 207

EXERCISE 14.2
Deciphering Plate Movements 209

EXERCISE 14.3
Anatomy of Subduction Zones 214

Chapter 15
Putting It All Together 218

EXERCISE 15.1
Deciphering the Tectonic Evolution of North America 220

EXERCISE 15.2
Relations between Tectonic Activity and Erosional Processes 223

EXERCISE 15.3
Geology Closer to Home 225

credits 228

PREFACE

Geology is best studied in its natural habitat—outdoors—where the rocks and minerals form, and the landscape is carved by streams, glaciers, and other agents of erosion. This is not always practical, and if the students cannot be brought to the field areas, the next best thing is to bring the field areas (via maps and photographs) and samples of their rocks and minerals to the students, in the laboratory. This manual is designed for use by students in the laboratories of introductory geology courses. It is not meant as a physical geology textbook, in that it does not try to present a comprehensive treatment of theories. Indeed, little explanation, description, or theoretical basis is given for the materials and processes covered by the exercises in this book, and there are few terms to memorize.

Instead, this manual contains a series of exercises designed to help the students discover for themselves what those explanations, underlying principles, and theories must be. The philosophy here is simple: it is more instructive (and more enjoyable) to figure out these principles for oneself than to memorize what other people have figured out. After all, geology is to a great extent a visual science, one whose processes can be observed and whose materials (minerals and rocks, water, air) can be seen, felt, tasted, and smelled. Many of the most important principles can be figured out with only some gentle prodding in the right direction. The stepwise nature of science—developing one idea from analysis of data and then using that idea to develop another concept—is what makes geology exciting and intellectually stimulating. If I have written this manual correctly, that excitement and intellectual challenge will be communicated to the students who use it.

This manual is divided into three main parts, each dealing with a different set of topics normally covered in introductory geology laboratories. A common theme runs through all three parts: how can particular materials, processes, and landscapes be used to reconstruct the history of the Earth? Thus, the goal of every chapter is to use geologic materials and features to construct hypotheses, not just to memorize the names of landforms or the hardnesses of minerals; to develop reasoning to the point that things as different as minerals, igneous rock textures, and glacial landforms can be used by students to describe something of the history of the mineral, rock, or area that they studied.

Section I introduces the nature of scientific reasoning and quickly applies that reasoning to the study of minerals (Chapter 1). Facts and skills concerning minerals and their identification are then applied to the study of rocks. Exercises show how igneous, sedimentary, and metamorphic rocks form, and how to recognize which processes were responsible for creating an individual rock (Chapter 2). These more advanced facts and skills are then applied to the study of all three kinds of rocks (Chapters 3, 4, and 5). When Part I is completed, students will be able to look at any rock and at least begin to describe its history.

Section II shows how different landscape features can be used to identify the particular agents of erosion that acted in a given area. This is done with a combination of color photographs of representative landforms and topographic maps of classic areas. Principles of map reading are introduced first (Chapter 6), including a series of exercises in which students teach themselves the "rules" of contour maps. Stream erosion and deposition exercises (Chapter 7) focus on the relation between how a stream does its geologic work and how that work is reflected by the surrounding landforms. Subsequent chapters (Chapters 8–11) on groundwater, glaciers, arid regions, and shorelines follow a similar format. They also teach by comparison: comparison of photographs and maps to determine, for example, the differences between stream-carved and glaciated valleys, or between landforms in arid and humid regions. Each chapter thus reinforces the others and helps students to synthesize and generalize while learning basic principles.

Section III focuses on large-scale features and how to interpret internal Earth processes (such as plate tectonics) from their effects on the surface. Students are first introduced to features resulting from deformation of rocks (Chapter 12), and are led through some simple steps to a point at which they are interpreting regional history based on the style and orientation of structures. A series of exercises in Chapter 13 shows how to interpret the sequence of events; then, portions of the geologic map of the United States are used for practice in doing just that on the grand scale. One exercise even visits the Moon to apply the same principles and reasoning. Chapter 14 leads students through some methods for determining the rate of large-scale plate tectonic processes, the directions of plate motion, and the geometry of subduction zones—from the physiography available on any good map, along with just a little bit of geologic information.

Chapter 15 asks the student to tie all of the concepts and skills together to interpret the geologic history of the area they call home (or of their campus). It is basically a series of questions designed to lead students progressively more deeply into the geologic evolution of the area in which they are learning geology for the first time.

To the student: Geology is not a cookbook-like experience in which all you have to do is follow a recipe (or instructions in a laboratory exercise) to come up with an explanation or a correct answer. Geology, like any science, is a series of discoveries, of hard work leading to the collection of data, of attempts to put all the information together to explain the things you see around you. This manual is meant to share with you the enjoyment of figuring out (geological) problems, of learning for yourself how things work. You will develop skills that will enable you to travel anywhere in the world and explain to your companions how the landscape around you was formed; to decide whether the land you are thinking of buying for your new home is suitable or not; to more thoroughly enjoy any vacation because you will know the origins of features around you.

To the instructor: If I have done my job correctly, your job should be easier than it has been with other laboratory manuals. Your students will be guided through a series of thought processes that more actively engage their brainpower and their interest. You will not have to lecture much during the laboratory session; lectures are replaced to a great extent by exercises in which the students learn for themselves what, in the past, you tried to get them to memorize. This does not mean that your job is at risk; there are still plenty of concepts that must be introduced, discussions to be led and coordinated with the course materials, and (I hope) many more intelligent questions to be answered.

The methods and individual exercises described in this manual have been tested over the past decade by graduate students at Queens College who serve as laboratory instructors in our introductory physical geology course. The exercises may be somewhat different in approach from other laboratory manuals, but our experience shows that they work. More exercises are contained in this manual than can be done in a typical semester or quarter. This is done so that you may customize your course to your specific needs and still be able to apply the philosophy, techniques, and selected exercises contained in the manual.

Finally, I would like to acknowledge the following people for their loan of sample materials: Dr. Sue Eriksson, VPI & SU, Blacksburg, VA; Dr. Alex Speer and Dr. Vic Cavaroc, NCSU, Raleigh, NC; Dr. Charles Almy, Guilford College, Greensboro, NC; and Ed Van Der Lippe of the Natural Science Center, Inc., Greensboro, NC.

Minerals and Rocks

Section 1

The history of the Earth is truly "written in stone." Every rock is a record of the processes by which it was made; many rocks also provide indications of the exact conditions in which those processes operated. Fossils in some rocks tell us the depth of the seas in which the ancient creatures lived, the climate on ancient continents, and even the age of the rock-forming process. Minerals in others help us estimate the high temperatures and intense pressures within the Earth; relations between these minerals reveal details of the sequence involved in development of the rocks.

The first five chapters of this manual are designed to help you learn how to read the history of the Earth, to open your eyes to the things you can deduce about the evolution of the rocks around you. Chapter 1 will introduce you to the world of minerals—the basic building blocks from which rocks are made. If rocks are the "books" in which Earth history is written, minerals are the words that, when put together in different ways, tell the fascinating story of the development of our planet through time.

Chapter 2 begins to help you see the different ways in which several minerals can be brought together in a rock—indeed to see how minerals grow in three different ways. You will learn the wide range of processes involved in creating minerals and the ways to determine which processes were involved in the formation of any rock.

Using minerals to unravel Earth history requires a keen eye and the mind of a detective. Some students think that learning to identify minerals is the main goal of their laboratory exercises, but that is just the beginning. A geological Sherlock Holmes asks not just, "What is this mineral?" but also, "How did it get here? What are its relations with the minerals around it?" Chapters 3, 4, and 5 will teach you some of the tricks of the trade used by geologists to read the different "languages" of the three kinds of rock.

CHAPTER 1

MINERALS

PURPOSE

To understand the differences between types of earth materials and to learn the logic and methods needed to identify individual minerals. To put these skills into use by identifying common rock-forming and economically valuable minerals.

MATERIALS NEEDED

1) A collection of earth materials
2) Supplies and equipment for determination of physical properties
3) A set of samples of common minerals

Introduction

Imagine an octopus (a highly intelligent creature) swimming along the sea floor when a shipload of footwear on the way to your local department store sinks and breaks open. Shoes, sneakers, boots, slippers, and sandals rain down around our octopus-scientist. What will it think of these things? After all, it doesn't walk, and neither it nor any of its acquaintances have feet. Our octopus wouldn't be able to recognize these foreign objects or to imagine what their use might be. Worse still, what do the terms "pair" or "left and right" mean to our eight-tentacled investigator?

What is an octopus doing in this physical geology laboratory exercise? You are about to begin the study of earth materials and processes, many of which may be as unfamiliar to you as sneakers are to a deep-sea creature. You may thus have the same problems that our hypothetical octopus did—the same problems that confront any scientist who is beginning to study something. Where do we begin? How can we make sense of things that we have never seen before? How can we understand the origins of unfamiliar objects or the causes of processes we have never before seen in action?

We start by making detailed observations and describing the objects as thoroughly as we can. We look for similarities and differences, and begin our scientific analysis by **classifying** the objects into groups whose members have something in common with one another. We ask how they are similar and how they differ from objects in other groups, and try to explain how the differences and similarities arose. In so doing, we begin the slow process of understanding something that had been totally unfamiliar. This is standard scientific procedure, so it is appropriate for us to begin your introduction to earth materials with a simple exercise in classification.

Exercise 1.1 Classification of Earth Materials

a) Examine the specimens of earth materials provided by your instructor. Which are similar to others? Which are different? Group the specimens into as many or as few categories as you feel are justified by your examination.

b) What criteria did you use in your classification? What was it about these features that led you to use them in separating the specimens into groups?

c) Our terminology for earth materials includes the words "mineral" and "rock." A **mineral** is a naturally occurring inorganic, homogeneous solid with an ordered internal arrangement of atoms and a characteristic chemical composition. A **rock** is an aggregate of grains of different minerals or of many grains of a single type of mineral.

d) Based on these definitions, which of the specimens are minerals and which are rocks?

Exercise 1.2 Physical Properties of Minerals

a) Examine one of the specimens that is composed of more than one kind of material. (This, because it is not homogeneous, must be a rock, and the different grains of which it is made must be minerals.) How many different minerals make up this particular rock?

b) Describe each of these minerals in your own words:

Mineral #1:

Mineral #2:

Mineral #3:

Mineral #4:

c) What aspects of each mineral's appearance enabled you to recognize that it was different from its neighbors? (A physical attribute that permits you to recognize a mineral is called a **physical property.**)

Introduction to the Physical Properties of Minerals

Approximately 3,000 minerals exist in nature. They differ from one another because each has a specific chemical composition and a unique three-dimensional arrangement of atoms within its structure. These differences result in a variety of physical properties, including the minerals' appearance, how they break, how well they resist being scratched, even how they smell, taste, and feel. You used some of these properties instinctively a moment ago to determine how many minerals were present in your specimen; others require a little explanation before they can be used.

Not all of these properties are equally useful. What if you were going to an airport to pick up someone you had never met, armed with a description provided by someone who had last seen your arriving passenger 25 years ago? Which features in the description would be most helpful? Height? Weight? Hair color? Some aspects of human features change markedly with time, while others, like eye color or shape of head, do not.

The problem is the same with minerals. Some properties never change. These are the most useful for identifying a mineral and are called **diagnostic properties.** Others, like a person's weight, may vary widely (not with time, but from specimen to specimen of the same mineral). You should be able to decide which of the following properties are truly diagnostic properties, and which are less useful.

Luster The luster of a mineral is the way its surface reflects light. Most terms used to describe luster are self-explanatory: metallic, earthy, waxy, greasy, vitreous (glassy), adamantine (or brilliant, as in a faceted diamond). It will be necessary, at least at first, only to distinguish between minerals with a metallic luster and those with one of the non-metallic lusters. A **metallic luster** is a shiny, opaque appearance similar to a bright chrome bumper on an automobile. Other shiny but somewhat translucent or transparent lusters (glassy, adamantine), along with dull, earthy, waxy, and resinous lusters, are grouped as **non-metallic.**

Color The color of a mineral is one of its most obvious attributes, and is one of the properties that must be given in any description. Color results from a mineral's chemical composition, impurities that may be present, and flaws or damage in the internal structure. Unfortunately, even though color is the easiest physical property to determine, it is not the most useful in helping to characterize a particular mineral. The problem is shown in figure 1.1, in which varieties of the mineral quartz display a rainbow of colors.

Is color a diagnostic physical property for quartz? Explain.

(Note that whether or not color is diagnostic for quartz, it may be diagnostic for other minerals.)

Streak The color of a mineral when it is powdered is called the **streak** of the mineral. Crushing and powdering a mineral eliminates some of the effects of impurities and structural flaws, and is therefore more diagnostic for some minerals than their color. Streak can be determined for any mineral by crushing it with a hammer, but it is more commonly (and less destructively) obtained by rubbing the mineral across the surface of a hard, unglazed porcelain material called a *streak plate.* The color of the powder left behind on the streak plate is the mineral's streak.

The streak and color of some minerals are the same. For others, the streak may be quite different from the color (Figure 1.2), most commonly for minerals with a metallic luster. The combination of luster, color, and streak may be enough to permit identification of the mineral.

Which property is more commonly diagnostic for a mineral—streak or color? Explain.

FIGURE 1.1 Specimens showing the variety of colors exhibited by the mineral quartz.

FIGURE 1.2 The difference between streak and color. For pyrite, on the right, the color of the mineral is different from the color of its streak. For cinnabar, on the left, however, the two are the same.

Crystal Shape When minerals form in environments where they can grow without interference from neighboring grains, they commonly develop into regular geometric shapes, called **crystals,** bounded by smooth crystal faces. The crystal form for a given mineral is governed by the mineral's internal structure, and may be distinctive enough to help identify the mineral (Figure 1.3). For example, quartz forms elongated, six-sided prisms capped with pyramid-like faces; galena and halite occur as cubes; and garnets develop 12- or 24-sided equidimensional forms.

Interference from other mineral grains during growth may prevent formation of well-formed crystals. The result is shapeless masses or specimens that developed only a few smooth crystal faces. This type of specimen is much more common than well-formed crystals.

Hardness Several items have been provided for your use in determining physical properties, including a glass plate, an unglazed porcelain streak plate, and a nail. Try scratching each of these with the others to determine their relative hardnesses. Where does your fingernail fit into this hardness scale? A copper penny?

Softest ——————————→ **Hardest**

The hardness of any object is controlled by the strength of bonds between atoms and is measured by the ease or difficulty with which it can be scratched. Diamond is the hardest mineral, because it can scratch all others. Talc is one of the softest; nearly every other mineral can scratch it. We measure a mineral's hardness by comparing it to the hardnesses of a standardized set of minerals first established by Friederich Mohs in the early nineteenth century, or with the common testing materials that have been calibrated to those standards (Figure 1.4).

The Mohs Hardness Scale is a *relative* scale. This means that a mineral will scratch any substance lower on the scale and will be scratched by any substance with a higher number. Diamond is not 10 times harder than talc or 1.1 times harder than corundum, as would be the case with an *absolute* hardness scale. Most often we are able only to narrow down hardness to within a certain range; for example, if an unknown mineral scratches a copper penny but does not scratch a glass plate, its hardness must be greater than 3.0 and less than 5.5.

Determine the hardnesses of the specimens indicated by your instructor as precisely as possible with your testing materials.

a) Galena, combination of cube and octohedron crystals

b) Pyritohedra of pyrite

c) Hexagonal crystals of beryl

d) Scalenohedra of calcite

e) Stubby crystals of wulfenite

f) Prisms of Vesuvianite

FIGURE 1.3 Differences in crystal shape indicate differences in the internal structure of minerals. Crystal shape may be a diagnostic physical property, as in these specimens.

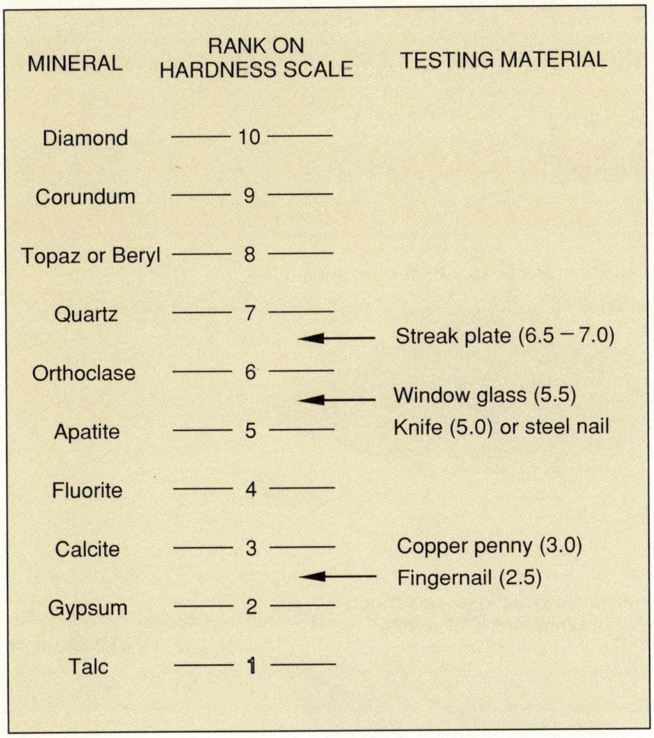

FIGURE 1.4 Mohs Scale of hardness compared with common testing materials.

FIGURE 1.5 Conchoidal fracture. This specimen of obsidian shows the type of fracture common in quartz and some other minerals.

Breakage The way in which a mineral breaks is controlled by the presence or absence of zones of weakness within its internal structure. This could be determined by hitting unknowns sharply with a small hammer, but that would destroy the samples. Instead, we carefully examine the specimen to see how it has already been broken—either by natural processes or when it was collected. Two different types of breakage are important: fracture and cleavage.

Fracture When bonds between atoms are approximately the same in all directions within a mineral, breakage occurs either on irregular surfaces (*splintery* or *irregular fracture*) or along smooth, curved surfaces (*conchoidal fracture;* Figure 1.5), similar to those formed when thick pieces of glass are broken.

Cleavage In some minerals, bonds between layers of atoms aligned in certain directions are weaker than bonds between different layers. In these cases, breakage occurs along smooth, flat surfaces parallel to those zones of weakness. In some minerals, a single direction of weakness exists, but in others, two, three, four, or as many as six (Figure 1.6) may be present. Where more than one direction of cleavage is present, it is important to determine the angular relation between the resulting cleavage surfaces: are they perpendicular to each other, or do they meet at an acute or obtuse angle?

When a mineral cleaves, it often exhibits many cleavage surfaces, but most of these are generally parallel to one another. A hundred cleavage surfaces parallel to one another all define a single **direction** of cleavage, because all of them are parallel to the same zone of bond weakness. It is the *number of directions* of cleavage that we record, along with the angles between them. Minerals with two or more cleavage directions generally have a "stair-step" appearance when viewed with a magnifying glass, as shown in Figure 1.6.

A mineral with two directions of cleavage may indeed be broken in some other direction—by irregular fracture. Thus, a single specimen may exhibit smooth cleavage planes in some directions, and irregular breakage surfaces in others. The more breakage surfaces we can see, the more clues we have to the mineral's internal structure.

It may be difficult for the beginning geologist to distinguish between cleavage and crystal faces. After all, both are smooth, planar surfaces. Two hints will help make the distinction easy. (1) If a mineral's outer surface shows a tarnish or alteration, the *crystal faces* will be tarnished or dull; if cleavage planes are present, they are usually recently made and will be fresher and less altered. (2) If many surfaces are present parallel to one another, they are most likely cleavage surfaces.

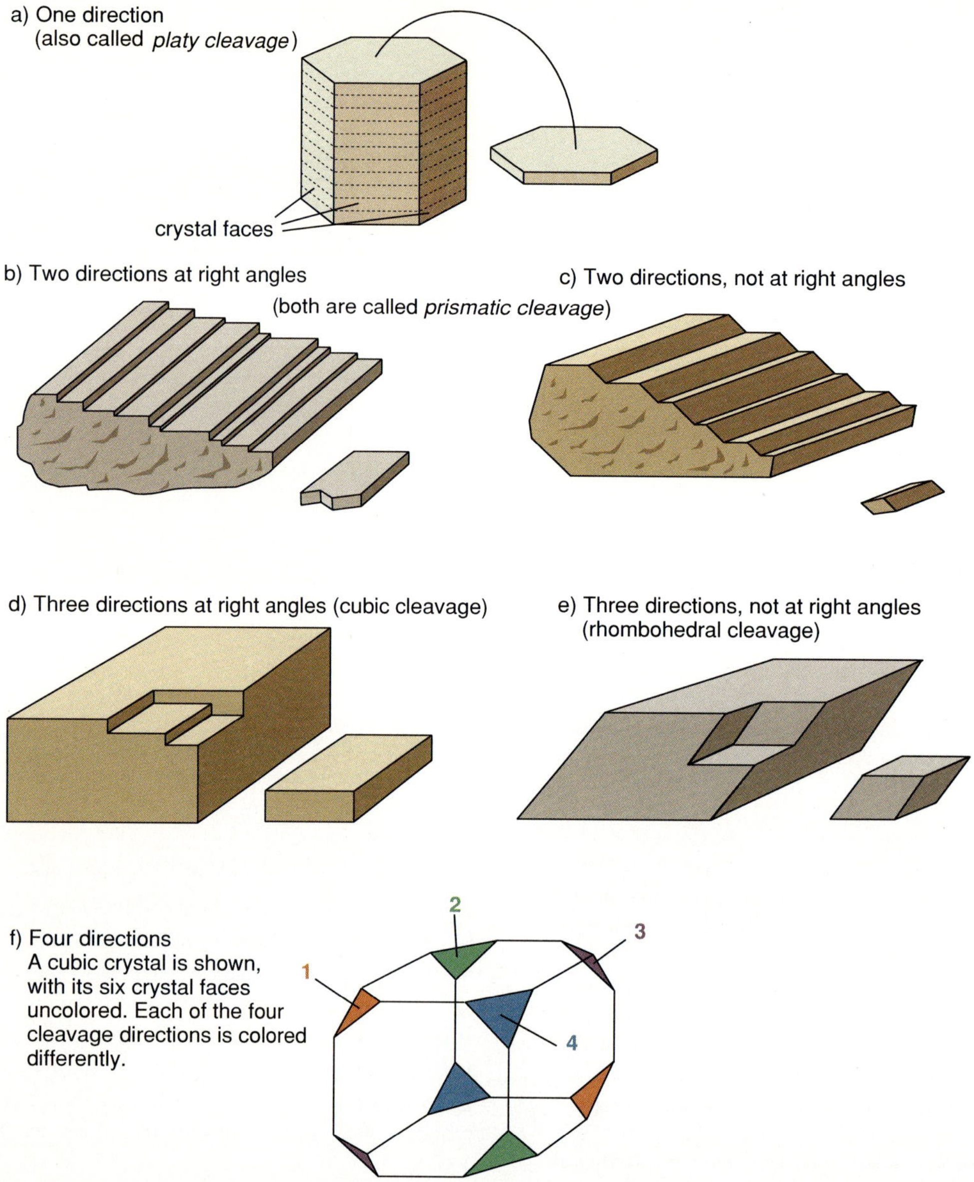

FIGURE 1.6 Types of cleavage. Directions of cleavage are color-coded.

Specific Gravity The specific gravity of a substance is a comparison of its density to that of water. Imagine a gallon bottle filled with water, a second filled with feathers, a third filled with lead weights. There are equal volumes of material present, but the bottle with the feathers will weigh less than that containing water; the bottle with lead weights will weigh the most. In order of increasing specific gravity, these materials would be: feathers, water, lead.

Specific gravity can be measured precisely, or estimated by a comparison, as above. To compare the specific gravity of any two minerals, simply hold a sample of one in your hand and "heft it," i.e., get a feeling for its weight. Then heft a sample of the other that is approximately the same size. If there is a great difference in specific gravity, you will detect it easily. It is often sufficient to note whether a mineral's specific gravity is significantly higher or lower than that of typical minerals.

Heft each of the specimens in your mineral set. Which ones have a high specific gravity? A low specific gravity?

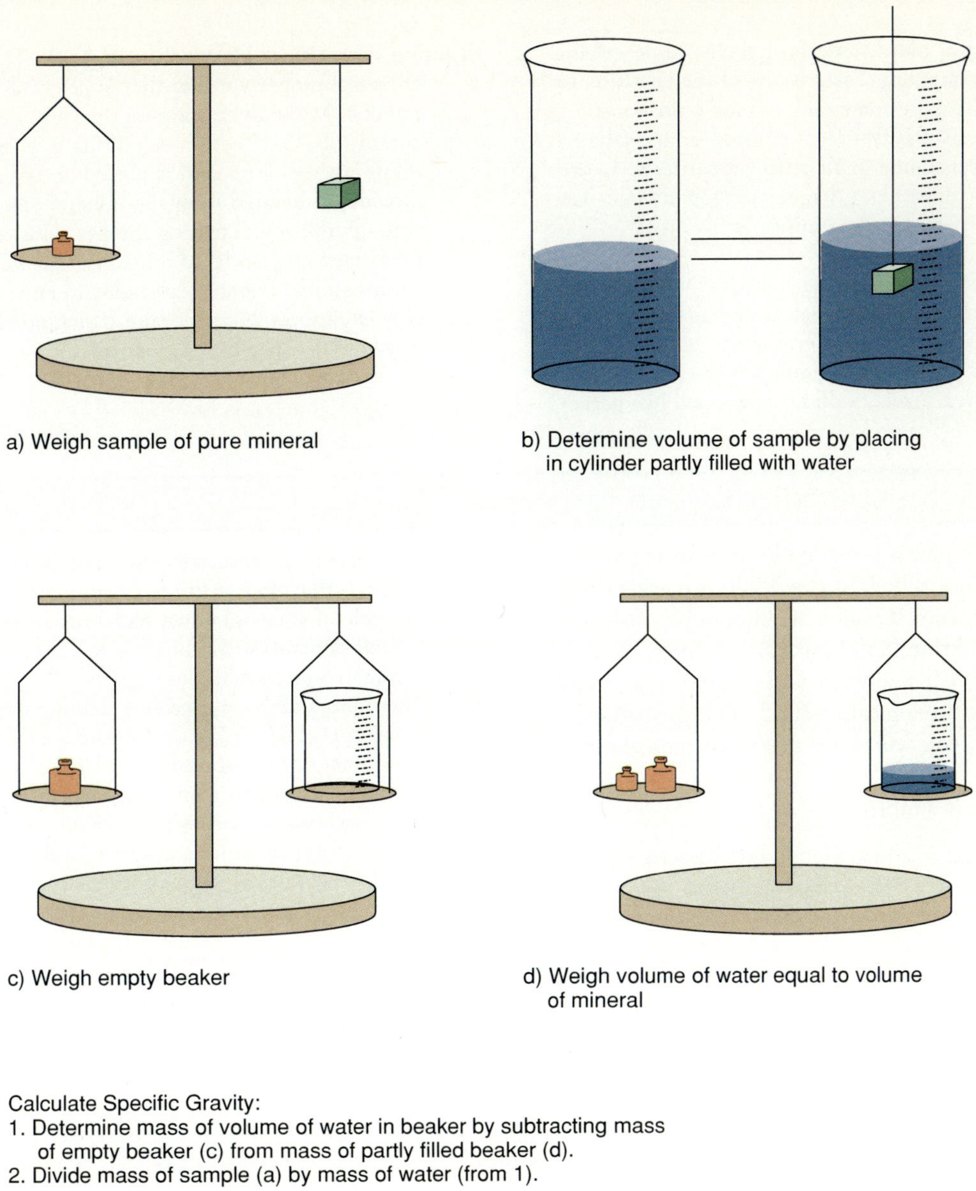

FIGURE 1.7 Steps in measuring specific gravity.

To measure the exact specific gravity of a mineral, it is necessary to compare the mineral's weight to the weight of exactly the same volume of water. Figure 1.7 shows how this can be done easily. First, weigh the mineral on a balance. Then, partially fill a graduated cylinder with water, and record the volume. Gently place the mineral in the cylinder and record the level of the water. The difference between the two readings is the volume of the specimen. Empty the cylinder, and fill a container with exactly the measured volume of water. Weigh that volume of water (being sure to subtract the weight of the dry container). Divide the weight of the specimen by the weight of the equal volume of water to calculate the specific gravity.

Magnetism A few minerals are attracted to a magnet or are themselves capable of acting as magnets (the most common magnetic mineral is magnetite). Because these are so rare, this property helps narrow the possibilities drastically when trying to identify an unknown specimen.

Feel Some minerals, notably talc and graphite, feel greasy or slippery when you rub your fingers over them. The greasiness occurs because bonds are so weak in one direction in these minerals that your finger pressure alone is enough to break them and to slide planes of atoms past neighboring atomic layers.

Taste Geologists use as many senses as possible in describing and identifying minerals. Taste is one of the last tests to be conducted, because some minerals are poisonous. Some minerals taste salty—most notably halite (salt). Sylvite, a mineral similar in all other properties to halite, tastes bitter. Taste is thus a diagnostic property because it distinguishes between these minerals.

Odor Odor? Yes, a few minerals have a distinctive smell; for others the powder left on a streak plate has the aroma. Sulfide minerals (such as sphalerite—zinc sulfide) may smell like rotten eggs. Arsenic minerals (such as arsenopyrite—iron arsenic sulfide) may smell like garlic!

Reaction with Dilute Hydrochloric Acid This is actually a chemical property rather than a physical attribute of a mineral. Minerals containing the carbonate anion complex $(CO_3)^{2-}$ effervesce ("fizz") when a drop of dilute hydrochloric acid is placed on them. Carbon dioxide is liberated from the mineral and bubbles out through the acid, creating the fizz. This test is best performed on powdered minerals. Calcite (calcium carbonate) will effervesce readily in either massive or powdered form, but dolomite (calcium-magnesium carbonate) reacts best as a powder.

EXERCISE 1.3 IDENTIFICATION OF MINERALS

Examine the set of minerals provided by your instructor. It is your job to determine which of the 3,000 minerals these specimens represent. No, it is not an impossible task, because (1) only a small number of minerals are important to us in our study of earth materials and mineral resources, and (2) if you follow the logical method of elimination described in figure 1.8, you will arrive at the correct identification.

The procedure is simple:

1) Study one mineral at a time and observe, determine, or measure as many of its physical properties as you can. Compile this information in the data sheets provided at the end of the chapter.

2) Use specific properties to eliminate from consideration minerals that do not fit your determination. For example, if your unknown had a hardness greater than 5.5, all minerals with hardness less than 5.5 are eliminated as possibilities.

3) These properties are used in a definite sequence—such as the one given in Figure 1.8—to steadily whittle down the number of possibilities, until (ideally) a single mineral remains. In many cases, only a few minerals will remain whose properties make them possibilities; these will be found in tables 1.1 and 1.2, along with more details that will help you choose among these possibilities. Congratulations; you have identified your unknown!

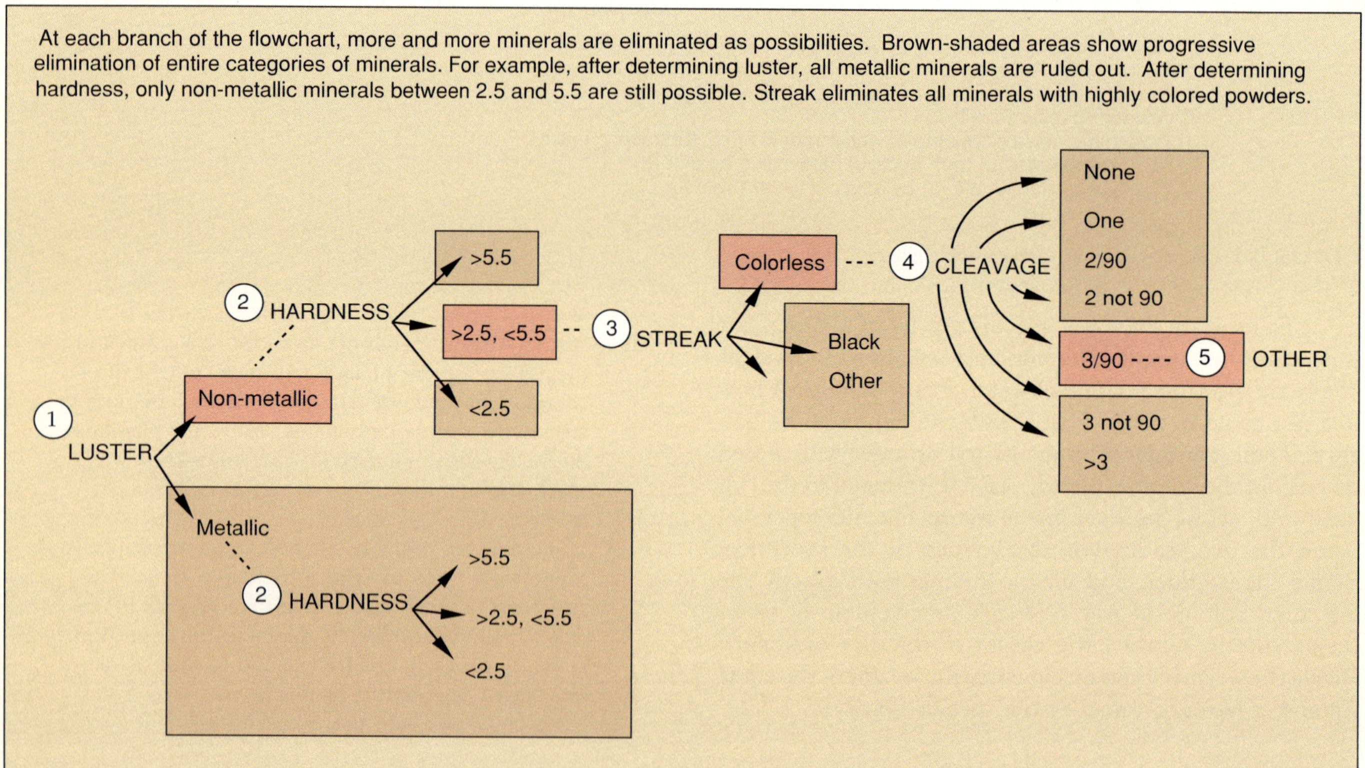

FIGURE 1.8 A flowchart for the systematic identification of minerals.

Table 1.1 Minerals with Metallic Luster

1. Hardness less than 2.5 (can be scratched by a fingernail)

STREAK	CLEAVAGE/FRACTURE	SP. GRAV.	OTHER	MINERAL (COMPOSITION)
Black	Perfect cleavage in one direction	2.23	Dark gray-black. Greasy feel. Leaves a mark on paper. H = 1	Graphite C
Black	One direction	4.62–4.73	Dark gray. Greasy feel. Often occurs in platy masses. H = 1.5	Molybdenite MoS_2
Black	One direction	4.52–4.62	Gray-black. Often in elongate, striated crystals. H = 2	Stibnite Sb_2S_3
Gray	3 directions at right angles	7.4–7.6	Lead gray color; commonly in cubic crystals. H = 2.5	Galena PbS
Red	3 directions, not at right angles▼	8.10	Bright, deep red. May also have non-metallic luster. H = 2.5	Cinnabar HgS
Yellow-Brown	▼	3.6–4.0▼	Occasionally in masses with metallic luster. More commonly as earthy masses.	Limonite $FeO(OH)\cdot n(H_2O)$

2. Hardness between 2.5 and 5.5 (harder than a fingernail, softer than glass)

STREAK	CLEAVAGE/FRACTURE	SP. GRAV.	OTHER	MINERAL (COMPOSITION)
Black	▼	4.1–4.3	Brassy yellow, often tarnished Similar to pyrite, but not in cubic crystals. H = 3.5–4.0	Chalcopyrite $CuFeS_2$
Black	▼	5.06–5.08	Bronze-brown when fresh; commonly with purplish, iridescent tarnish. H = 3	Bornite Cu_5FeS_4
Black	▼	4.58–4.65	Brown-bronze color. Magnetic to varying degrees. H = 4.0	Pyrrhotite $Fe_{1-x}S$
Black	▼	4.6	Submetallic luster; brownish-black; rare 8-sided crystals. H = 5.5	Chromite $FeCr_2O_4$
Copper-Red	▼	8.9	Copper color; often found in branchlike masses. Rare 4-, 8-, and 12-sided crystals. H = 2.5–3.0	Copper Cu
Yellow	▼	19.3	Golden yellow; malleable; rare 8-sided crystals. H = 2.5–3.0	Gold Au
Yellow	6 directions; only a few are usually present in a single specimen	3.9–4.1	Most are resinous. Only a few varieties have a submetallic luster. Brown, yellow, black and colorless varieties. H = 3.5	Sphalerite ZnS

▼This property is rarely seen or measurable.

W A R N I N G

Minerals do not always read this book to find out what their properties should be. Variations in growth, degree of weathering, chemical impurities, and many other phenomena may result in a mineral displaying properties other than those listed here. Tables 1.1 and 1.2 will work for you in most cases, however. (The only sure way to identify an unknown mineral is by determining its chemical composition and internal atomic structure. Since this requires years of training and expensive equipment, we will do the best we can with the simple physical properties.)

TABLE 1.1 MINERALS WITH METALLIC LUSTER (continued)

3. HARDNESS GREATER THAN 5.5 (HARDER THAN A KNIFE)

STREAK	CLEAVAGE/FRACTURE	SP. GRAV.	OTHER	MINERAL (COMPOSITION)
Black	Conchoidal fracture	5.02	Brassy yellow; commonly in cubes or 12-sided crystals with striated faces. H = 5	Pyrite FeS_2
	▼	5.18	Iron black; strongly magnetic. H = 6	Magnetite Fe_3O_4
	▼	6.07	Silver-white; streak smells like garlic. H = 5.5–6.0	Arsenopyrite FeAsS
	▼	4.7	Iron-black. Some specimens may be faintly magnetic. H = 5.5–6.0	Ilmenite $FeTiO_3$
Red-Brown	▼	5.26	Only the rare black variety has a metallic luster and H = 5.5–6.5	Hematite Fe_2O_3

▼This property is rarely seen or measurable.

TABLE 1.2 MINERALS WITH NON-METALLIC LUSTER

1. HARDNESS LESS THAN 2.5 (SOFTER THAN A FINGERNAIL)

STREAK	CLEAVAGE/ FRACTURE	SP. GRAV.	OTHER	MINERAL (COMPOSITION)
Orange	1 direction▼	3.48	Red-orange; resinous luster Streak may smell like garlic. H = 1.5–2.0	Realgar AsS
Yellow	1 direction▼	3.49	Lemon-yellow; resinous luster Streak may smell like garlic. H = 1.5–2.0	Orpiment As_2S_3
	Conchoidal to uneven fracture	2.05–2.09	Various shades of yellow; resinous luster. H = 1.5–2.5	Sulfur S
White or Colorless	Perfect cleavage in 1 direction	2.76–2.88	A colorless mica; can be peeled into transparent sheets. H = 2–2.5	Muscovite $KAl_2(AlSi_3O_{10})(OH)_2$
	Perfect cleavage in 1 direction	2.6–3.3	Various shades of green Mica-like mineral. H = 2–2.5	Chlorite Complex Fe-Mg sheet silicate
	Perfect cleavage in 1 direction	2.7–2.8	Greasy feel; green, gray, and white varieties common. H = 1	Talc $Mg_3Si_4O_{10}(OH)_2$
	Perfect cleavage in 1 direction (clear type) Fibrous fracture	2.32	Clear, colorless, and fibrous or massive white and gray varieties. H = 2	Gypsum $CaSO_4 \cdot 2H_2O$
	3 directions at right angles	1.99	Colorless or white; cubic crystals. Tastes bitter. H = 2	Sylvite KCl
	Perfect cleavage in 1 direction▼	2.6	Usually occurs in dull, earthy, powdery white masses that stick to the tongue. H = 2–2.5	Kaolinite $Al_2Si_2O_5(OH)_{10}$

▼This property is rarely seen or measurable.

TABLE 1.2 MINERALS WITH NON-METALLIC LUSTER (continued)

1. HARDNESS LESS THAN 2.5 (SOFTER THAN A FINGERNAIL)

STREAK	CLEAVAGE/ FRACTURE	SP. GRAV.	OTHER	MINERAL (COMPOSITION)
White or Colorless	Perfect cleavage in 1 direction (platy type)	2.5–2.6	Platy and fibrous (asbestos) varieties. Greasy luster, generally green color. H = 2–5	Serpentine $Mg_3Si_2O_5(OH)_{10}$
	▼	2.0–2.55	Not really a mineral; rock containing several Al-rich clay minerals, often in small spherical particles.	Bauxite Mixture of Al-hydroxides
Green or Brown	Perfect cleavage in 1 direction	2.8–3.2	A dark-colored mica, usually brown or green. Can be peeled into thin sheets. H = 2.5–3.0	Biotite $K(Fe,Mg)_3(AlSi_3O_{10})(OH)_2$

2. HARDNESS BETWEEN 2.5 AND 5.5 (HARDER THAN A FINGERNAIL, SOFTER THAN A KNIFE)

STREAK	CLEAVAGE/ FRACTURE	SP. GRAV.	OTHER	MINERAL (COMPOSITION)
Green		3.9–4.03	Bright green; often occurs in globular or elongate masses. Reacts with HCl.	Malachite $Cu_2CO_3(OH)_2$
Blue		3.77	Intense blue; often occurs in platy crystals or in radiating spherical masses. Reacts with HCl. H = 3.5–4.0	Azurite $Cu_3(CO_3)_2(OH)_2$
Red	3 directions, not at right angles▼	8.10	Adamantine to earthy masses and rare crystals. Bright red. H = 2.5	Cinnabar HgS
Red-brown		5.26	Reddish-brown; often in earthy masses. H = 5.5–6.5	Hematite Fe_2O_3
Yellow-Brown		3.6–4.0▼	Commonly as earthy, powdery masses, and as coating on other minerals.	Limonite $FeO(OH)\ n(H_2O)$
	3 directions, not at right angles	3.96	Light to dark brown; often in rhombic crystals. Reacts with hot HCl. H = 3.5–4.0	Siderite $FeCO_3$
	6 directions▼	3.9–4.1	Resinous luster; colors include white, colorless, black, brown. H = 3.5	Sphalerite ZnS
White or Colorless	3 directions at right angles	2.16	Colorless to white; cubic crystals. Salty taste. H = 2.5	Halite NaCl
	3 directions, not at right angles	2.71	Varicolored; usually white or colorless. Rhombic or elongate crystals. Reacts with HCl. H = 3	Calcite $CaCO_3$
	3 directions, not at right angles	2.85	Varicolored: often pink or white Rhombic crystals. Powder reacts with HCl. H = 3.5–4.0	Dolomite $CaMg(CO_3)_2$
	3 directions at right angles	4.5	Colorless or white. Unusually high Sp.G. for nonmetallic mineral. H = 3–3.5	Barite $BaSO_4$
	4 directions	3.18	Cubic crystals; varicolored-light green, purple, blue, yellow, colorless. H = 4	Fluorite CaF_2
	1 direction, poor	3.15–3.20	Elongate 6-sided crystals Usually green or brown; rarely purple, blue, colorless. H = 5	Apatite $Ca_5(PO_4)_3(OH,F,Cl)$

▼This property is rarely seen or measurable.

TABLE 1.2 MINERALS WITH NON-METALLIC LUSTER (continued)

3. HARDNESS BETWEEN 5.5 AND 7.5 (HARDER THAN A KNIFE, SOFTER THAN STREAK PLATE)

STREAK	CLEAVAGE/ FRACTURE	SP. GRAV.	OTHER	MINERAL (COMPOSITION)
Brownish Black	2 directions at right angles▼	4.18–4.24	Reddish brown to black; often has brilliant luster. H = 6–6.5	Rutile TiO_2
White or Colorless	Conchoidal fracture	3.5–4.3	Equidimensional 12-sided crystals; varicolored—red, green, brown, black. H = 7	Garnet family Complex Ca,Fe,Mg,Al,Cr,Mn silicate
	2 directions at right angles	2.54–2.62	Three polymorphs: orthoclase, microcline, sanidine; colorless, pink, gray, green. H = 6	Potassic feldspar $KAlSi_3O_8$
	2 directions at right angles	2.62–2.76	Solid-solution series between Albite (Na) and Anorthite (An) Colorless, white, gray, black. Striations on one cleavage face. H = 6	Plagioclase feldspar $CaAl_2Si_2O_8$ $NaAlSi_3O_8$
	Conchoidal fracture	2.65	Elongate, 6-sided prisms. Wide range of color: colorless, pink, purple, black, yellow, green. H = 7	Quartz SiO_2
	▼	3.16–3.20	Elongate, 4-sided crystals; gray, white, red, brown, green. H = 7.5	Andalusite Al_2SiO_5
	1 direction of cleavage	3.23	Long, slender crystals; often as fibers. Usually white, but may be green. H = 6–7	Sillimanite Al_2SiO_5
	1 direction of cleavage	3.55–3.66	Bladed crystals, usually blue or white. H = 7 parallel to length of crystal, 5 at right angles.	Kyanite Al_2SiO_5
Colorless to Pale Green	Conchoidal fracture	3.27–4.37	Solid solution between Fayalite (Fe) and Forsterite (Mg). Green stubby crystals or granular masses in rocks. H = 6.5–7.0	Olivine $(Fe,Mg)_2SiO_4$
	2 directions at 56 and 124°	3.0–3.4	An amphibole with double chain structure. Dark green to black prismatic crystals. H = 5–6	Hornblende Chain silicate with Ca,Na,Fe,Mg, Al,OH
	2 directions at 56 and 124°	3.0–3.3	An amphibole with double chain structure. Elongate pale to dark green crystals. H = 5–6	Actinolite Chain silicate with Ca,Fe,Mg,OH
	2 directions at 87 and 93°	3.2–3.3	A pyroxene with single chain structure; dark green to black prismatic crystals. H = 5–6	Augite Chain silicate with Ca,Na,Mg,Fe,Al
	1 perfect, 1 poor Not at 90°	3.25–3.45	Apple green to black color Elongate crystals, some striated. H = 6–7	Epidote Complex silicate with Ca,Al,Fe,OH

▼This property is rarely seen or measurable.

Table 1.2 MINERALS WITH NON-METALLIC LUSTER (continued)

4. HARDNESS GREATER THAN 7.5 (HARDER THAN STREAK PLATE)

STREAK	CLEAVAGE/ FRACTURE	SP. GRAV.	OTHER	MINERAL (COMPOSITION)
Mineral scratches streak plate; does not leave powder	1 direction	3.4–3.6	Elongate crystals with complex pointed ends. Colorless, blue, white, green, yellow, pink. H = 8	Topaz $Al_2SiO_5(F,OH)_2$
	1 direction (imperfect)	2.65–2.8	Elongate 6-sided crystals with flat ends. Blue-green, emerald green, yellow, pink, white. H = 8	Beryl $Be_3Al_2(Si_6O_{18})$
	2 directions at right angles (poor)	4.68	Brilliant luster; commonly brown but colorless, green, red types also occur. H = 7.5	Zircon $ZrSiO_4$
		4.02	6-sided, barrel-shaped crystals Blue (sapphire), red (ruby), brown, pink, white, green. H = 9	Corundum Al_2O_3
	4 directions▼	3.51	Brilliant luster; equidimensional 8-sided crystals. Colorless, blue, pink, yellow. H = 10	Diamond C

▼This property is rarely seen or measurable.

MINERAL IDENTIFICATION DATA SHEET

NAME ____________________

SAMPLE NUMBER	LUSTER	HARDNESS	STREAK	COLOR	CLEAVAGE OR FRACTURE	OTHER PROPERTIES	MINERAL NAME AND COMPOSITION

MINERAL IDENTIFICATION DATA SHEET

NAME ______________________

SAMPLE NUMBER	LUSTER	HARDNESS	STREAK	COLOR	CLEAVAGE OR FRACTURE	OTHER PROPERTIES	MINERAL NAME AND COMPOSITION

Chapter 2

Minerals in Rocks

Purpose

To understand the kinds of relations between grains in different kinds of rock, and to use those relations to identify the processes by which any given rock was created. To practice identification of minerals as they occur in rocks, rather than as individual crystals or homogeneous masses.

Materials Needed

1) A collection of igneous, sedimentary, and metamorphic rocks.
2) Supplies and equipment for making artificial igneous, sedimentary, and metamorphic rocks in the classroom.
3) A magnifying glass or hand lens.

INTRODUCTION

A rock is an aggregate of mineral or organic grains brought together by the igneous, sedimentary, or metamorphic processes discussed in your textbook. Most rocks contain several different minerals, some just two or three minerals, and a few are composed of many grains of a single mineral. The following exercises will help you identify minerals when they occur with others in a rock, and introduce you to the importance of texture in interpreting a rock's history. You will also see the processes by which some of the most common textures form.

Exercise 2.1 IDENTIFICATION OF MINERALS IN A ROCK

Identification of minerals in a rock should follow the same procedures used in identifying individual minerals, but there are some potential traps that you should be aware of:

When trying to determine hardness or streak, be sure you are testing only a single mineral grain. Make sure that your knife or nail is scratching a single mineral grain or crystal (i.e., the knife is harder than the mineral), rather than dislodging it from the rest of the rock (which teaches you only that you can break the rock apart easily!).

The specific gravity of a single mineral cannot be measured if it is present with several others in a rock, or if there are pore spaces visible in the rock.

Rocks often develop a weathered outer surface that is different in color from the fresh surface. Be sure to describe both, because the combination of weathered and fresh colors is often a valuable clue as to the minerals present in fine-grained rocks. The weathered color is not necessarily the true color of any of the minerals present.

The color of a rock often depends on its grain size. All other things being equal, a fine-grained rock will appear darker in color than a coarse-grained rock made up of the same minerals.

Be careful when applying hydrochloric acid! Look with your magnifying glass to learn exactly which minerals are effervescing. Are all of the grains reacting with the acid (in which case all are calcite), or are some of them nonreactive (in which case not everything is a carbonate)?

Most of the rocks of the Earth's crust are made from a surprisingly small number of minerals, most of which should already be familiar to you from Exercise 1.3.

Briefly examine the rock specimens provided. Which contain "old friends?" Which of the minerals that you studied in Exercise 1.3 are present?

Specimen # ___ *Minerals present*

____

____

____

____

____

____

____

____

Exercise 2.2 A ROCK IS MORE THAN THE SUM OF ITS MINERALS

In some cases, a rock's mineralogy is enough to tell what type of process was involved in its formation. A rock composed entirely of halite, for example, is a sedimentary rock, because halite forms only by precipitation from saltwater. A rock containing muscovite, biotite, garnet, and staurolite must be metamorphic, because only metamorphism can produce this assemblage. Table 2.1 lists some of the minerals that are most useful in determining the nature of the rock-forming process.

In many cases, mineralogy alone is not enough to determine a rock's origin. The most common rocks are composed of minerals that can be produced by two or more

TABLE 2.1 OCCURRENCE OF THE MOST COMMON ROCK-FORMING MINERALS			
MINERAL	IGNEOUS	SEDIMENTARY	METAMORPHIC
Plagioclase Feldspar	▼		▼
Potassic Feldspar	▼		▼
Quartz	▼	▼	▼
Hornblende	▼		▼
Actinolite			▼
Augite	▼		
Muscovite	▼		▼
Biotite	▼		▼
Chlorite			▼
Olivine	▼		▼
Garnet	▼		▼
Andalusite			▼
Kyanite			▼
Sillimanite			▼
Epidote			▼
Halite		▼	
Calcite		▼	▼
Dolomite		▼	▼
Gypsum		▼	
Talc			▼
Serpentine			▼

▼Most commonly grown during these rock-forming processes

rock-forming processes. Quartz, potassic feldspar, and sodium-rich plagioclase feldspar can all form by igneous, metamorphic, **or** sedimentary processes. A rock made of these three minerals could thus belong to any of the three categories. In these cases, we must look at the texture of the rock, as well as its mineralogy, to determine its origin.

The **texture** of a rock includes such features as the sizes and shapes of its grains, and the manner in which the grains are related to one another. Just as you learned how to use physical properties to identify a mineral, you will now see how to use the texture and mineralogy of a rock to interpret its origin.

a) Carefully examine the rock samples provided. How many different minerals are present in each rock? What minerals are present in each? (Hint: the same physical properties that you used to identify individual minerals should be useful here.)

	Rock # __	*Rock # __*	*Rock # __*
Minerals			

b) Write a brief description of each rock. A complete description should include the rock's appearance (color or combination of colors), mineralogy, and density. Try your hand at describing the **textures.** Do the grains interlock as a three-dimensional jigsaw puzzle, or are individual grains separated from one another by smaller ones without interlocking? Are grains angular or rounded? Are they coarse (>5 mm) or fine?

Rock # __

Rock # __

Rock # __

c) What features do these rocks have in common? What features are different?

d) Why is texture an important attribute of a rock?

e) Are these rocks probably the products of the same kind of rock-forming process? (Note that you can arrive at a reasonable answer to this question without knowing very much about how each rock formed!)

EXERCISE 2.3 THE ORIGIN OF TYPICAL ROCK TEXTURES

a) Place a glass petri dish on a hot plate and put in enough of the substance provided to cover the bottom. Raise the heat until this substance melts completely. Remove the dish (carefully; it's hot) from the hot plate and allow it to cool. Congratulations! You have just made an igneous rock that displays a typical igneous texture.

1) Sketch and describe the textures present.

2) Describe the crystallization process. How did individual grains seem to grow, and how did they attain their final relation with their neighbors? (This is an **interlocking texture,** which is characteristic of most igneous rocks, but may be found in sedimentary and metamorphic rocks as well. In sedimentary and metamorphic rocks exhibiting interlocking texture, the identities of the minerals present will determine the rock-forming process.)

3) Compare this texture with those of the rocks in Exercise 2.1. Which of those rocks most likely formed by igneous processes?

---

b) Repeat this procedure with another petri dish partly filled with sugar.

1) Sketch and describe the textures present.

2) Describe the crystallization process. How did individual grains seem to grow, and how did they attain their final relation with their neighbors?

3) This is a much different texture found in some igneous rocks, called a **glassy texture.** Suggest an explanation for the formation of the two different textures.

c) Cover the bottom of a glass dish with a mixture of sand grains and small pebbles. Add some of the liquid provided and allow the mixture to sit until most of the liquid has evaporated. Now turn the glass upside down. You have made a model of a sedimentary rock called a **clastic rock:** composed of individual grains that were physically weathered from older rocks, transported, deposited, and converted to solid rock again. This is a **clastic texture.**

1) Why don't the grains fall out of the dish?

2) Examine the "rock" with your magnifying glass. Sketch and describe the texture.

3) Compare this "rock" with those used in Exercise 2.1. Which of those rocks have a clastic sedimentary history?

d) Take a handful of plastic chips and push them into the mass of Play-doh, with a random orientation. Then flatten the Play-doh against the tabletop with your palm. You have just replicated a texture found commonly in metamorphic rocks that have experienced intense directed pressures.

1) Sketch and describe the orientation of the flat chips in your "metamorphic rock." Are they randomly oriented as they were when you started? Explain their present orientation.

2) The parallel alignment of flat grains in a metamorphic rock is known as a **foliated** texture. What does such a texture tell you about the metamorphic process that produced the rock?

3) Compare this texture with that of the rocks used in Exercise 2.1. Which of those rocks (if any) exhibits this metamorphic foliated texture?

Exercise 2.4 Identification of Rocks as Igneous, Sedimentary, or Metamorphic

Examine all the other rocks in your set. Separate them into igneous, sedimentary, and metamorphic categories based on their textures and mineral contents, using the flowchart in figure 2.1 and information from table 2.1.

A fourth category—"don't know yet"—is perfectly reasonable in some instances, particularly for the fine-grained rocks. Your instructor will give you clues to sort out the very fine-grained rocks. Geologists can mount slices of rock on glass slides and grind the rock so thin that you can see through it, then use a special microscope to study the rock. If a petrographic microscope and thin sections of the fine-grained rocks are available, you may be able to identify the textures even though they cannot be seen with the naked eye.

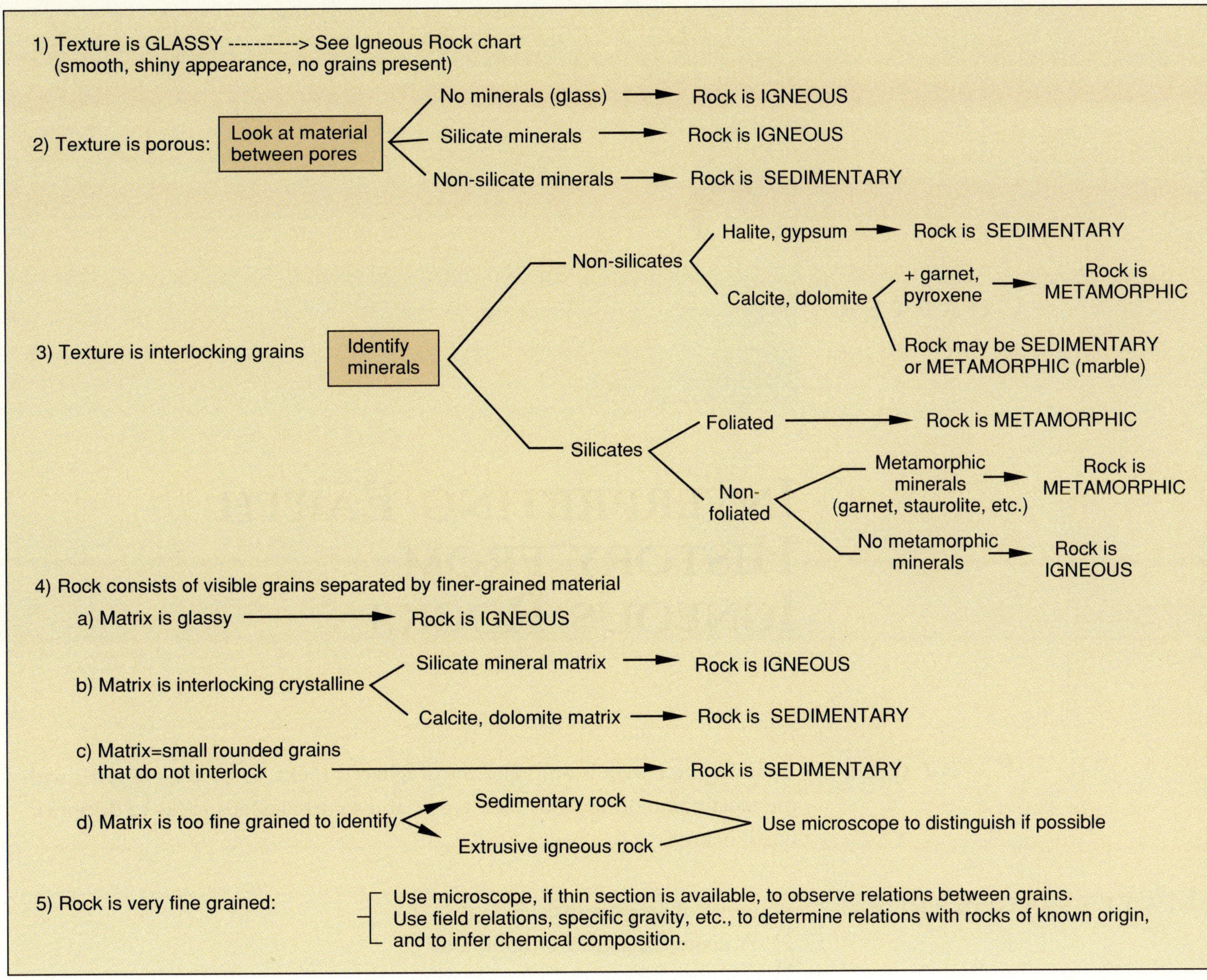

FIGURE 2.1 Flowchart for rock classification.

Probably Igneous	Probably Sedimentary	Probably Metamorphic	Don't Know Yet
________	________	________	________
________	________	________	________
________	________	________	________
________	________	________	________
________	________	________	________
________	________	________	________
________	________	________	________
________	________	________	________
________	________	________	________

Chapter 3

Interpreting Earth History from Igneous Rocks

Purpose

To learn the minerals and textures of igneous rocks, to become familiar with the most common igneous rock types, and to use igneous rocks to decipher Earth history.

Materials Needed

1) A set of igneous rocks
2) A magnifying glass or hand lens
3) Supplies for mineral identification

Introduction

Igneous rocks are those formed by the cooling and solidification of molten rock material called **magma.** Magma sometimes solidifies within the Earth, producing what are called **intrusive** (or **plutonic**) igneous rocks. We would never know about these if it were not for processes of erosion that remove the overlying material and eventually expose the intrusive rock. For example, the Sierra Nevada Mountains of California, including Yosemite National Park, are largely underlain by intrusive rocks. When magma reaches the surface, it is called **lava,** and the rocks it produces are called **extrusive** (or **volcanic**) rocks. Such rocks formed the state of Hawaii and make up large portions of California, Oregon, Washington, and Idaho.

Exercise 3.1 The Importance of Texture in Studying Igneous Rocks

Separate either all the light-colored igneous rocks **or** all the dark-colored igneous rocks from your study set. Write a description for each of these rocks, being sure to include: color (fresh and altered), minerals present, and textural features such as grain size (coarse or fine, uniform or non-uniform), grain shape, and relations between grains.

Specimen #	**Description**
____________	____
____________	____
____________	____
____________	____
____________	____

It is entirely possible that all of these rocks formed from exactly the same magma—that is, all have precisely the same chemical composition. What characteristics of the processes that formed these rocks could have led to the differences observed?

These rocks look so different from one another that each has a different name, even though all may be the same chemically and mineralogically.

Origin of Igneous Textures

Two factors determine what an igneous rock will look like. The chemical composition of the magma determines what minerals can crystallize, but the **cooling history** of the magma controls the actual appearance of the rock. The texture of an igneous rock reveals its unique cooling history and permits you to determine whether the rock is intrusive or extrusive.

In Exercises 2.3a and 2.3b, you saw how molten material could cool to form two very different igneous textures. There, we used different materials to take advantage of their different melting points and structural complexity. An even greater degree of textural variation can occur in nature from a single magma, depending on how it cooled. The two critical factors are (1) the rate at which the magma cooled, and (2) the presence or absence of gas in the magma.

Rate of Cooling When a magma starts to cool, a few small seed crystals form throughout the magma. Atoms move toward them, attach to them, and cause the crystals to grow. The more slowly a magma cools, the more time its atoms and ions have to move toward the seeds. The more time there is for this process, the larger the crystals can become. If cooling is rapid, however, atoms cannot reach the original seeds before the magma solidifies; instead, many other seeds form, but these only have enough time to add atoms from their immediate neighborhood. The result is many small crystals.

Role of Gases Shake a can of carbonated beverage and pop the top. The gas escapes, carrying with it some of the beverage as a frothy, foamy liquid. When large amounts of gas escape from a cooling magma, a similar foam forms, which may solidify if the magma cools quickly. The violent release of gases during some eruptions blasts spatters of magma and blocks of solid rock into the air. These fall to Earth and form fragmental rocks called **tuffs.**

Exercise 3.2 Interpreting Igneous Textures

a) Which of the specimens that you studied in Exercise 3.1 cooled quickly? Which ones cooled slowly?

____

b) Which of the specimens in your igneous rock set cooled from a magma rich in gases?

c) Some of your samples have non-uniform grain sizes. Based on cooling rate alone, how can this happen?

Examine one of these specimens. Which minerals crystallized early in the cooling history of this rock? How do you know?

d) One (or more) of your specimens looks like a chunk of glass, with no mineral grains at all. Suggest an explanation for the origin of this rock.

e) Which type of cooling (rapid or slow) would you expect to find in intrusive magmas? Extrusive lavas? Explain.

f) Which of the specimens in your rock set are intrusive, and which are extrusive?

Intrusive **Extrusive**

Notice that it is now possible for you to deduce much about the cooling history of an igneous rock, without knowing either the name of the rock or the geological terms for the textural features you are using as evidence. Table 3.1 describes the texture, provides the correct terms, and summarizes the interpretations that each textural feature permits. Figure 3.1 shows examples of each of the textural features.

TABLE 3.1 INTERPRETATION OF IGNEOUS TEXTURES

TEXTURE	DESCRIPTION	INTERPRETATION	EXAMPLE
Pegmatitic	Very coarse grained (>2.5 cm)	Cooling from an extremely fluid magma; usually intrusive	Figure 3.1a
Coarse grained (phaneritic)	Individual grains visible to the naked eye	Slow cooling; generally intrusive	Figure 3.1b
Fine grained (aphanitic)	Individual grains not visible to the naked eye	Rapid cooling; generally extrusive	Figure 3.1c
Porphyritic	A few large grains (called phenocrysts) set in a much finer grained matrix	Two cooling rates: slow at first to form phenocrysts, then rapid to crystallize the matrix	Figure 3.1d
Glassy	Smooth, shiny; looks like glass No mineral grains present	Extremely rapid cooling; generally extrusive	Figure 3.1e
Porous (vesicular)	Spongy; full of large or small holes	Rapid cooling accompanied by the escape of gases	Figure 3.1f
Pyroclastic	Fragmental; pieces set in a fine-grained matrix	Violent, explosive eruption	Figure 3.1g

FIGURE 3.1 Textures of igneous rocks.

e) Glassy texture

f) Vesicular texture (pumice)

Vesicular texture (scoria)

g) Pyroclastic texture

FIGURE 3.1 continued

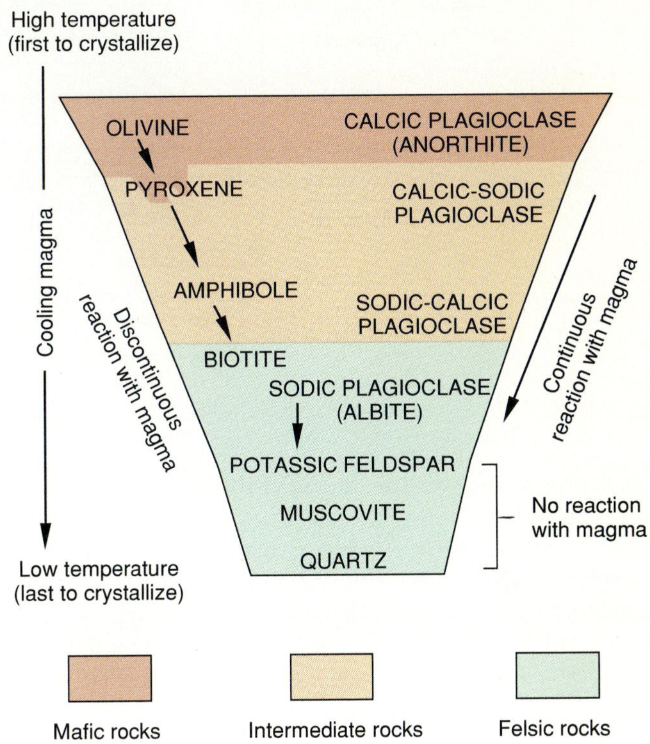

FIGURE 3.2 Relation between Bowen's Reaction Series and the major groups of igneous rocks.

IDENTIFICATION OF IGNEOUS ROCKS

With three thousand possible minerals, and cooling processes that can produce many different textures, there might at first seem to be an endless number of different igneous rocks. Relax. Just as most of the Earth's crust is composed of only a few minerals, these minerals combine in only a few ways in igneous rocks. The reason is partially explained by Bowen's Reaction Series (Figure 3.2), which was originally intended to show the sequence of mineral crystallization from magma.

Magmas produce rocks containing minerals that crystallize at approximately the same temperature. Thus, minerals at the top of Bowen's Reaction Series crystallize at high temperatures and are commonly found together in igneous rocks. Intermediate and low temperatures have their own characteristic assemblages. The three resulting rock types are called **mafic, intermediate,** and **felsic** igneous rocks. *Mafic* refers to the high proportion of *ma*gnesium and iron (*f*errum in Latin) minerals, such as olivine and pyroxene. *Felsic* refers to the predominance of potassic *fel*dspar and sodium-rich plagioclase *fel*dspar. Intermediate rocks, as you would expect, have compositions intermediate between these two types. A fourth group of rocks is called **ultramafic.** These contain *only* the iron-magnesium minerals, with very little, if any, feldspar.

Figure 3.3 is a simplified classification scheme for the most common igneous rocks based on their mineralogy and texture. To identify an igneous rock, use its mineral content to decide which of the four rock groups it belongs to. Then move downward through the diagram until you reach the

LIGHT COLORED → DARK COLORED

INCREASING SPECIFIC GRAVITY →

	FELSIC	INTER-MEDIATE	MAFIC	ULTRA-MAFIC
MINERALOGY	Quartz Potassic feldspar Sodium plagioclase +/- Biotite, hornblende muscovite	Ca/Na Plag. Hornblende +/- Augite	Ca-Plag. Augite +/- Olivine	Pyroxene +/- Olivine
TEXTURE ** Pegmatitic	Granitic pegmatite		Mafic pegmatite	
Coarse grained	Granite	Diorite	Gabbro	*
Fine grained	Rhyolite	Andesite	Basalt	Very rare rock types
Glassy	Obsidian		Tachylite	
Vesicular	Pumice		Scoria	
Pyroclastic	Rhyolite tuff	Andesite tuff	Basalt tuff	

* Ultramafic rock name depends on mineral(s) present: Olivine only--Dunite. Olivine + pyroxene--Peridotite. Pyroxene only: Pyroxenite.

** See text for correct naming of porphyritic rocks.

FIGURE 3.3 Classification of igneous rocks.

line for the correct texture, and read the name assigned to this rock type.

Any rock with a porphyritic texture can be called a **porphyry,** but a more precise name indicates the composition and the grain size of the matrix. For example: a fine-grained felsic rock with a few large potassic feldspar crystals would be a **rhyolite** porphyry; a coarse-grained felsic rock with a few very large potassic feldspar crystals would be a ***granite*** porphyry.

What do you do with very fine-grained rocks whose minerals you can't even see, much less identify? Or with glassy rocks that have no minerals at all? (Both obsidian and tachylite look like black glass.) In these cases, the rock's color and specific gravity are very helpful. Mafic and ultramafic rocks are made up of dark-colored minerals with high specific gravities. It is not surprising that they are darker and denser than felsic rocks, which are composed of light-colored minerals with relatively low specific gravities.

Devise a simple test for distinguishing between samples of obsidian and tachylite.

EXERCISE 3.3 IDENTIFICATION OF IGNEOUS ROCKS AND INTERPRETATION OF THEIR COOLING HISTORIES

Determine the mineralogy and texture of each sample in your igneous rock set. Record your data on the sheets provided at the end of the chapter. Use figure 3.3 to determine the name for each rock, and indicate as much of the cooling history as you can. You should be able to write a brief paragraph for each specimen, detailing how and where it crystallized and, in some instances, the exact sequence of mineral growth.

DATA SHEET FOR IGNEOUS ROCKS

NAME ______________________

SPECIMEN NUMBER	MINERALS PRESENT (APPROXIMATE PERCENTAGES)	TEXTURE	NAME OF ROCK	COOLING HISTORY

CHAPTER 4

INTERPRETING EARTH HISTORY FROM SEDIMENTARY ROCKS

PURPOSE

To become familiar with the minerals and textures of the most common sedimentary rocks; to learn how to use sedimentary rocks to unravel the history of Earth's surface.

MATERIALS NEEDED

1) A set of sedimentary rocks
2) A magnifying glass or hand lens
3) Supplies for mineral identification
4) Supplies for making chemical sedimentary rocks

Introduction

Sedimentary rocks form at the Earth's surface by the physical and chemical breakdown of older rocks, and the recombination of the "pieces" into new arrangements. Because they form at the surface, sedimentary rocks provide us with clues to what it would have been like to live on the surface of the Earth millions or even billions of years ago. What was the climate in Chicago 500,000,000 years ago? How different were the shapes of oceans and locations of continents 1,000,000,000 years ago? Did dinosaurs really wander through Montana, Texas, New Jersey, Massachusetts, and Georgia in the past? If you know what to look for, sedimentary rocks can provide the answers to questions such as these.

Sedimentary rocks are by far the most abundant rock type found at the surface today. In many places, they form a blanket over the igneous and metamorphic rocks from which they are derived. For example, the Great Plains region of the midcontinent consists of sedimentary rocks eroded from the Rocky Mountains over millions of years; similarly, the Atlantic coastal plain is made of debris shed from the Appalachian Mountains.

Formation of Sedimentary Rocks

Figure 4.1 outlines the four steps required to make sedimentary rocks: **weathering** produces loose particles (sediment) from previously existing rocks; **transportation** moves those particles to an area where they can be collected; **deposition** causes the particles to settle to the bottom of the collecting basin; and **lithification** transforms the sediment into a cohesive, solid rock again. The mineralogy, texture, and a few structures unique to sedimentary rocks enable geologists to piece together this complex four-stage process.

There are two kinds of weathering. In **physical weathering,** a rock is broken into smaller pieces without changing any of the minerals that are present. This is accomplished by several mechanisms, including plant roots that enter cracks and then grow, wedging out chunks of the rock, and water that seeps into cracks, freezes, and expands. In **chemical weathering,** water, carbon dioxide, oxygen, and other gases in the atmosphere attack the minerals themselves and gradually dissolve them.

The different types of weathering produce two very different kinds of sediment. Physical weathering yields **clasts** that are transported physically by the agents of erosion, and chemical weathering yields **ions** that are transported chemically—by solution in water. A third kind of sedimentary particle is formed when living organisms use some of the ions to build shells, skeletons, or plant tissue. When the organisms die, their more resistant parts are transported physically, like mineral clasts.

Some sedimentary rocks are composed entirely of one of these types of sediment. These are called **clastic, chemical,** or **biogenic** rocks. In many instances, a rock is eventually formed by a combination of two or all three of the possible components. We shall see later how to name these.

Textures of Sedimentary Rocks

Relations between Grains

In Exercise 2.3c, you made a clastic sedimentary "rock" by cementing mineral grains together. Many clastic and biogenic rocks have this type of texture, in which mineral grains or shell fragments are cemented by minerals that have precipitated in pore spaces between the clasts. The most common natural cementing agents are calcite, quartz, hematite, and limonite. A rock's color is usually enough to help identify the last two as cements, and a well-placed drop of hydrochloric acid identifies calcite.

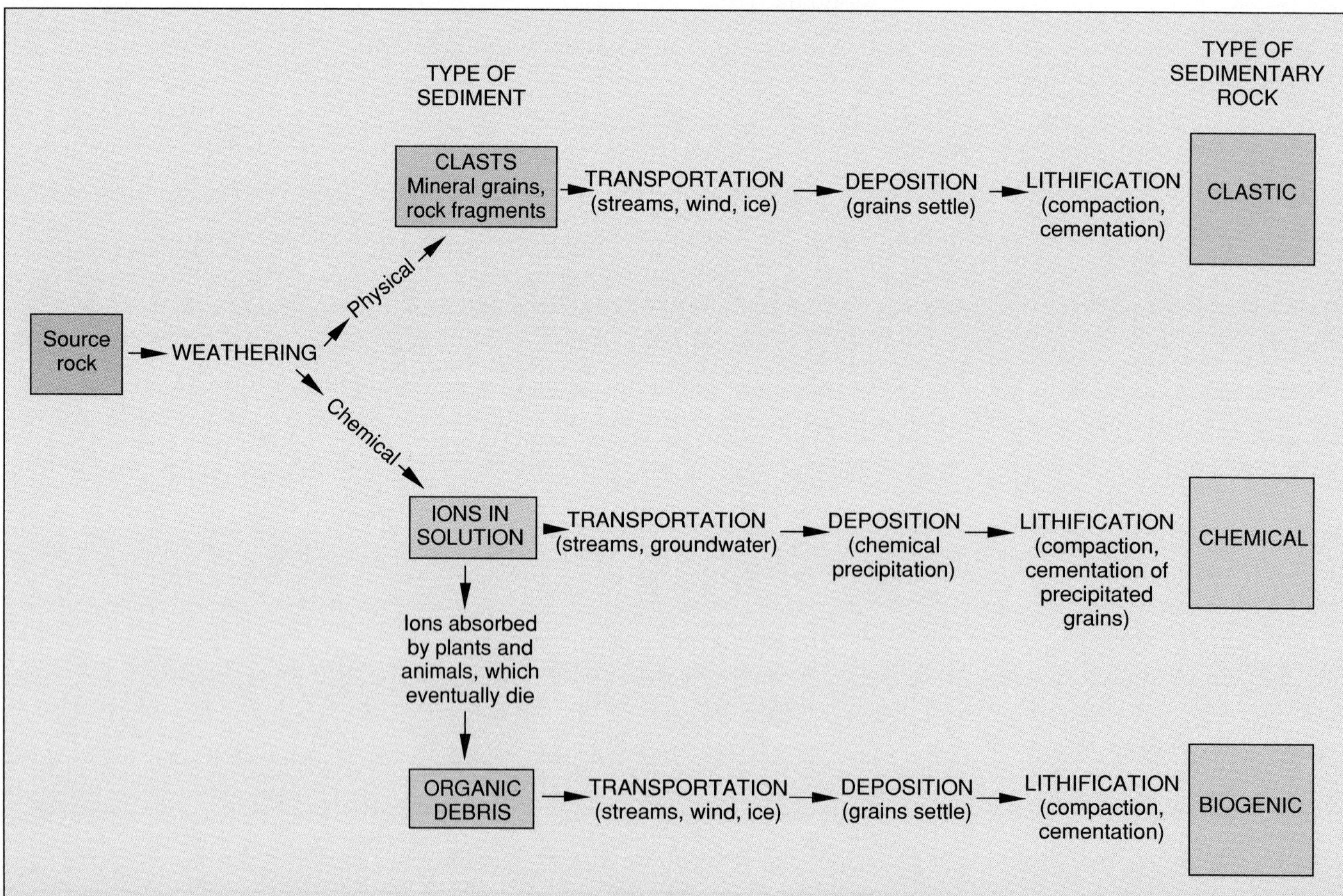

FIGURE 4.1 Steps in the formation of sedimentary rocks.

Which of the rocks in your set are clastic sedimentary rocks?

When cement is not present to hold the clasts together, lithification commonly occurs by a process called **compaction.** The pressure of overlying rocks squeezes water out of the loose sediment and produces cohesion. You can do this yourself (not in the laboratory, please) by picking up a handful of wet mud and squeezing hard—the result is a coherent mudball. Nature does an even better job; beneath the surface, pressures can become so great that some grains dissolve partially where they come into contact with others, and the dissolved mineral material immediately precipitates and coats the grains, binding them tightly together.

EXERCISE 4.1 TEXTURES OF CHEMICAL SEDIMENTARY ROCKS

Chemical sedimentary rocks may have two different textures, both of which can be duplicated in simple experiments.

---

a) Fill a beaker ¾ full with the colorless solution provided. Blow (gently) through a straw into the liquid until you notice a change in the contents of the beaker. **(Do not suck on the straw!)**

What change did you observe in the liquid?

Congratulations! You have just made some chemical sediment. Carbon dioxide from your lungs combined with calcium and oxygen ions from the solution to precipitate tiny grains of calcite. Now let the beaker sit for a few minutes so that gravity can take over and cause the calcite to be deposited on the bottom of the beaker. Gently pour off the liquid and examine the sediment.

Can you see individual grains with the naked eye? With your magnifying glass? With a microscope?

Many chemical sedimentary rocks have grains this size that have precipitated from seawater.

---

b) Place some seawater in a shallow dish, and set the dish on a hot plate until the water evaporates. (For those of you not conveniently located near an ocean, use the salt water provided.)

1) Examine the residue after evaporation is complete, preferably with the aid of a microscope or powerful magnifying glass. Describe what you see. How are these grains different from the chemical sediment that you made in Exercise 4.1a?

2) How do relations between these grains differ from those between the grains in a clastic rock?

Under appropriate conditions, minerals precipitated in chemical sedimentation have the same interlocking texture that you saw in igneous rocks. In most cases, however, the *minerals* are so different that you would never mistake the rock for an igneous rock. Common evaporite minerals include halite, gypsum, and anhydrite—these are never igneous.

Which of the samples in your rock set are probably chemical sedimentary rocks?

Grain Size

Grain size in an igneous rock depends primarily on one thing—the rate at which the magma cools—but the size of grains in a sedimentary rock depends on several factors, which are controlled by the kind of sediment involved. The grain size of biogenic rocks depends on the size of the organism involved and the degree to which breakage has occurred during transportation. Thus, the calcareous shells of microscopic organisms that make up **chalk** (see figure 4.3e) result in an extremely fine-grained rock, whereas the large shells or shell fragments of **coquina** (see figure 4.3f) yield a much coarser rock. Grain size in chemical rocks depends on the number of ions that migrate to crystal seeds during precipitation and growth.

Grain size is more useful in analyzing clastic rocks than chemical rocks. Clasts range from boulders bigger than a house to grains so small that they cannot be identified, even with a powerful microscope. The sizes of grains in a clastic rock depend on the sizes of the fragments created by physical weathering, and on how the particles were transported.

The more kinetic energy an erosional agent has, the bigger the particles it can carry. In terms of how much kinetic energy each has, glaciers are much more energetic than streams; streams are more energetic than wind. Extremely large grains (the house- or car-sized variety) can be moved only by a glacier. Streams can move particles up to the size of small boulders; wind is able to move grains not much larger than the period at the end of this sentence. If the largest grain present in a rock is larger than what a particular agent can carry, you have eliminated one of the possible transporting mechanisms.

Geologists assign specific size ranges to grain-size terms commonly used by laymen; they then name rocks based on their grain sizes, as shown in table 4.1. Figure 4.2 shows examples of these common clastic rocks.

You don't have to carry a ruler and microscope to distinguish the fine grain sizes. If you can see individual grains with your magnifying glass, the grains are at least sand-sized. If a rock is so fine grained that you can't see the grains, rub the rock across your front teeth (very gently). If it feels rough, the grains are silt sized; if smooth, the grains are mud-sized.

Sorting The degree to which grains in a clastic rock are uniform in size is called the rock's **sorting.** This property is even more useful in identifying the agent of transportation than grain size alone. Wind-transported particles are very well sorted, because wind can only pick up and move grains within the sand size range. Streams can move anything from mud to small boulders, but when streams deposit their clasts, they tend to do so gradually, sorting them moderately well. Because they are solid, glaciers carry everything together and deposit it all at once. As a result, glacial deposits are usually very poorly sorted. Sometimes, as during a flood, streams may lose energy rapidly, resulting in less well-sorted deposits.

Which, if any, of your specimens is potentially a glacial deposit?

TABLE 4.1 SIZES OF SEDIMENTARY PARTICLES AND NOMENCLATURE OF CLASTIC SEDIMENTARY ROCKS

SIZE TERM	SIZE RANGE	ROCK NAME
Boulder	>256 mm (>10 inches)	Boulder conglomerate
Cobble	64–256 mm (2.5–10 inches)	Cobble conglomerate
Pebble	4–64 mm (0.15–2.5 inches)	Pebble conglomerate
Granule	1–2 mm (0.04–.15 inches)	Granule conglomerate
Sand	0.06–2 mm	Sandstone
Silt	0.004–0.06 mm	Siltstone
Mud	<0.004 mm	Mudstone or shale

Grain Shape

The shape of grains, like grain size, is more useful for analyzing clastic rocks than chemical or biogenic types. The shapes of the shells and crystals control the shape of grains in biogenic and chemical rocks, but transportation and deposition help determine grain shape in clastic rocks. By examining the shapes of clasts, you can deduce the agent of transportation and the distance the particle traveled.

Roundness vs Angularity Particles carried in air or water are free to move about and, as a result, often collide with one another. The collisions knock off sharp corners and edges and eventually produce nearly spherical grains. In contrast, the grains in a glacier are frozen into position and experience none of these collisions. Glacially transported particles thus tend to be much more angular than those carried by wind or streams.

Oolitic Texture Some chemical precipitates occur initially as coatings on tiny clasts close to the shore where they are rolled back and forth by currents. Precipitation may continue as the grains roll, causing a buildup of concentric layers around the original grain and eventual formation of a small sphere, roughly the size of a salmon egg, called an oolite (or ooid) (Figure 4.3). The most common oolites are composed of calcite, quartz, or apatite.

a) Conglomerate

b) Breccia

c) Sandstone

d) Siltstone—smooth areas are imprints of a fossil plant.

e) Mudstone—the impressions have been turned over so that their lower surface can be seen

f) Shale

FIGURE 4.2 Common clastic rocks.

EXERCISE 4.2 INTERPRETATION OF TRANSPORT MECHANISM BASED ON TEXTURAL FEATURES

Based on the information given earlier and on your common sense, fill in the blanks in the following chart.

TRANSPORT MECHANISM			
	STREAM	WIND	GLACIER
Texture			
Grain Size Range			
Sorting			
Roundness			

MINERALOGY OF SEDIMENTARY ROCKS

You might expect sedimentary rocks to contain an enormous variety of minerals, because they are formed from pieces of all sorts of source rocks. However, chemical weathering greatly reduces the number of minerals found in clastic rocks by destroying many of the original igneous and metamorphic minerals. In general, minerals that form at high temperatures and pressures tend to weather rapidly at the Earth's surface, where it is cool and where the gases of the atmosphere can combine with water to attack the mineral structures. The ferromagnesian minerals, particularly olivine, pyroxenes, and amphiboles, are quickly converted to limonite and hematite in a humid environment. Feldspars alter to clay minerals such as kaolinite. Only the most resistant minerals survive to become clasts; quartz, potassic feldspar, and sodium-rich plagioclase are the most abundant. In general, minerals high on Bower's Reaction Series (Figure 3.2) are less resistant than those lower in the series.

Clay minerals, calcite, dolomite, halite, and gypsum are truly sedimentary minerals, forming during the surface processes that produce sedimentary rocks. The clays are transported along with the resistant mineral grains and are important constituents of clastic rocks. Calcite, dolomite, halite, and gypsum are most commonly found in chemical rocks.

a) Chert (hand specimen above and photomicrograph below)
b) Micritic limestone (hand specimen above and photomicrograph below)
c) Oolitic limestone (hand specimen above and photomicrograph below)
d) Rock salt
e) Chalk
f) Coquina
g) Coal
h) Fossiliferous limestone

FIGURE 4.3 Chemical and biogenic sedimentary rocks

Quartz is nearly ubiquitous in sedimentary rocks. It is the most abundant resistant mineral and is thus the principal component of most clastic rocks. Quartz is also an important component of chemical rocks. Chemical weathering of ferromagnesian minerals releases silicon and oxygen ions into the solution. If enough ions are concentrated in the solution, they recombine, and grains of quartz precipitate. This is how the chemical rock *chert* forms.

Sedimentary Structures

The environments in which clastic rocks were deposited are often revealed by features called **sedimentary structures** that formed during transportation or deposition. The most common sedimentary structure is a **bed,** a horizontal layer of sediment that represents a single depositional event (Figures 4.4a, b). A bed differs from those above and below it in ways that may be subtle (slight change in grain shape, size, or sorting) or not very subtle at all (for example, a change from green to red or black color, or from mudstone to boulder conglomerate).

Exercise 4.3 Graded Beds

Many beds are homogeneous from top to bottom, but in some cases there is a systematic change in grain size. Partially (about ¼ to ⅓) fill a glass cylinder or tall jar with a mixture of pebbles, sand, silt, and mud-sized grains, and add water to just below the top. Screw on the jar top or place your hand over the opening, and shake the container vigorously for a few seconds. Then stand it upright and watch what happens.

a) Describe in as much detail as you can what happens in the first few seconds of "deposition."

b) What happens after a few minutes have elapsed?

c) Sketch the bed that has formed from your experiment.

For what should now be obvious reasons, this type of bed is called a **graded bed.** It is extremely useful to geologists studying highly deformed rocks, because it can be used to tell which side of the bed was originally the top and which was originally the bottom.

d) Sketch what you think a graded bed would look like if it were overturned several million years after it formed.

Ripplemarks and Cross-bedding When sand grains are deposited in the presence of either air or water currents, they frequently develop a surface with undulating features, called ripplemarks, such as those shown in figure 4.4c. A sand dune, for example, is simply a large-scale ripple produced by the wind. In most cases, the direction of the air or water current is perpendicular to the crest of the ripplemark, as shown in the left photograph in figure 4.4c. Note that the ripplemark has a steep side, facing downcurrent (the direction **toward** which the current flowed), and a gentle side facing the upcurrent direction. These are called asymmetric ripplemarks. The right-hand photograph in figure 4.4c shows a symmetrical ripplemark. Note that both of its sides are equally steep.

FIGURE 4.4 Sedimentary structures.

e) Based on what you know about asymmetric ripplemarks, suggest an origin for the symmetrical types. (HINT: they are also called oscillation ripplemarks.)

---

f) Sketch a profile across a set of symmetrical and asymmetrical ripplemarks. Now turn the page upside-down. Which of the two types can be used, like graded beds, to determine where the top of a bed was?

As a ripple develops, grains roll up the gentle upcurrent side and drop down the downcurrent side. This produces a layering that is not horizontal, the way it is in most beds; this layering is called **cross-bedding.** Figure 4.4b shows a large-scale example of cross-bedding from Zion National Park in Utah.

Mudcracks and Raindrop Impressions Mix flakes of clay minerals with water to produce a thick slurry, and pour about 1/8″ of this mixture into a glass baking dish. Set the dish on a hot plate, and let it sit until the water starts to evaporate. Periodically touch the surface gently with a pencil eraser until the surface has hardened.

---

g) Describe and sketch the features that develop around the margins of the dish; does the hardened material fill the width and length of the dish as it did when it was liquid, or has it pulled away? Explain how this happened.

In the real world of sedimentary environments, the drying of a layer of mud commonly leads to mudcracks caused by the same process that you observed in this experiment (Figure 4.4d).

---

h) Imagine that instead of the end of a pencil, several large raindrops had fallen on the slowly hardening surface. Sketch what you think the result would be on the surface.

Look at figure 4.4e to see how good your imagination was.

Fossils A fossil is the remains or trace of a living organism found in a rock (Figure 4.4f). Fossils may come from animals or plants, and be as large as the thigh bone of a *Brontosaurus*, or as tiny as a microscopic spore from an ancient fern. Fossils may be the actual remains of an organism, the imprint of its body, or the footprints of a walking animal. Fossils enable us to interpret sedimentary environments in great detail, and permit us to determine the locations and boundaries of ancient oceans and lakes.

---

i) From figure 4.4f, what can you say about the environments in which the two organisms lived?

Identification and Interpretation of Sedimentary Rocks

Table 4.2 is a simplified classification scheme for sedimentary rocks. Use it in conjunction with the mineralogical and textural information you have learned to identify the sedimentary rocks in your set. Compile your data in the sheets provided at the end of the chapter.

You should be able to write a short paragraph about the history of each sample. That paragraph should include the type of weathering involved in producing the particles that now reside in your rock, the agent of transportation, and anything about the environment of deposition that you can deduce. As in all scientific endeavors, you must discuss the evidence that led to your conclusions.

Table 4.2 Classification of Sedimentary Rocks

A) CLASTIC ROCKS (INDIVIDUAL GRAINS SET IN CEMENT OR MATRIX)

GRAIN-SIZE TERM	GRAIN SIZE	HOW TO TELL	GRAIN SHAPE		COMPOSITIONAL MODIFIERS BASED ON CLAST MINERALOGY		
			ROUNDED	ANGULAR	QUARTZ	QTZ + FSPAR	ROCK FRAGMENTS
Gravel	>2 mm	Measure	Conglomerate	Breccia	Quartzose	Arkosic	Lithic
Sand	1/16–2 mm	Grains visible	Sandstone		Orthoquartzite	Arkosic	Graywacke
Silt	1/256–1/16 mm	Feels gritty on the teeth	Siltstone		Too fine grained to tell composition		
Mud	<1/256 mm	Feels smooth on the teeth	Mudstone or shale		Too fine grained to tell composition		

B) BIOGENIC ROCKS

TYPE OF ORGANISM	ROCK NAME
Mass of shell fragments	Coquina
Fine-grained mass of microscopic calcareous organisms	Chalk
Fine-grained mass of microscopic siliceous organisms	Diatomite
Plant debris	Peat; Coal

C) CHEMICAL ROCKS

MINERAL	TEXTURE	
	INTERLOCKING CRYSTALS	VERY FINE GRAINED
Halite	Rock salt	
Gypsum		Alabaster
Calcite	Crystalline limestone	Micritic limestone
Dolomite	Crystalline dolostone	Dolomicrite
Quartz		Chert (flint)

D) COMBINATIONS (EXAMPLES OF THE POSSIBILITIES)

Quartz-rich conglomerate with calcite cement and a few shell fragments: **Calcite-cemented fossiliferous quartzose conglomerate**

Very fine-grained dolostone with abundant quartz and feldspar grains: **Arkosic dolomicrite**

DATA SHEET FOR SEDIMENTARY ROCKS

NAME ____________

SPECIMEN NUMBER	TEXTURAL FEATURES (GRAIN SIZE, SHAPE, ETC.)	MINERALS PRESENT (APPROXIMATE PERCENTAGES)	ORIGIN OF ROCK	NAME OF ROCK

CHAPTER 5

INTERPRETING EARTH HISTORY FROM METAMORPHIC ROCKS

PURPOSE

To learn the minerals and mineral assemblages found in metamorphic rocks; to learn how different metamorphic textures form; to become familiar with the most common metamorphic rock types; and to use metamorphic rocks to decipher aspects of earth history.

MATERIALS NEEDED

1) A set of metamorphic rocks
2) A magnifying glass or hand lens
3) Mineral identification materials

Introduction

Metamorphism occurs within the Earth when previously existing rocks are subjected to conditions of heat, pressure, and chemical activity that are greatly different from those under which they first formed. The original rock may be sedimentary, igneous, or even an older metamorphic rock. Some of the original minerals break down in response to the new conditions, and new **metamorphic minerals** replace them. As a result, the original texture may also be obliterated, and replaced by a texture reflecting the processes of metamorphism involved.

As with intrusive igneous rocks, erosion strips off as much as thousands of meters of overlying strata to reveal metamorphic rocks that formed deep in the Earth's crust. Vast areas of metamorphic rock form the cores of every continent and are present in linear mountain belts, such as the Appalachians, Rockies, Himalayas, and Alps, that formed as the result of the collision of lithosphere plates. Metamorphic rocks record the conditions and processes that occurred within the Earth, and thus provide clues to the nature of parts of the crust that we have never seen. They also reveal the processes by which continents and mountains form.

The nature of a metamorphic rock is determined by a combination of three factors: (1) the composition of the original rock, (2) the type of metamorphism involved, and (3) the intensity of the metamorphism. Once you know what to look for, you will be able to find evidence for each of the three factors in your samples.

The Role of Composition

Exercise 5.1 Mineralogy of Metamorphic Rocks

a) Examine the four metamorphic rocks indicated by your instructor, and identify the minerals present in each. Record the chemical composition of each mineral (from memory, Table 1.1, or your textbook).

	SPECIMEN ___	SPECIMEN ___	SPECIMEN ___	SPECIMEN ___
M				
I				
N				
E				
R				
A				
L				
S				

b) Based on the chemical compositions of the minerals, which elements were present in the parent rock of each of these four specimens?

	SPECIMEN ___	SPECIMEN ___	SPECIMEN ___	SPECIMEN ___
Elements				

c) Did any of these specimens originate from the same parent rock? Explain.

In most cases, the atoms present in the parent rock are the only ones that metamorphism can rearrange to create new minerals. Thus, the original chemical composition controls what minerals can form. (The exceptional cases involve a type of metamorphism in which large numbers of atoms are either added to or removed from the original rock.)

We will consider four basic kinds of parent rock. All contain oxygen, the most abundant element in the crust; they differ in the proportions of the other common elements.

▼

Shales Rocks composed of clay minerals rich in aluminum, silicon, iron, magnesium, and potassium.

▼

Calcareous Rocks Limestones and dolostones rich in calcium, magnesium, carbon, and oxygen, with less iron and magnesium and only a little aluminum.

▼

Mafic Rocks Basalts and gabbros rich in calcium, aluminum, iron, and magnesium.

▼

Quartzo-Feldspathic Rocks Felsic igneous rocks like granite and rhyolite, or clastic sedimentary rocks such as arkose. These contain abundant silicon, potassium, aluminum, and sodium, with minor iron or magnesium.

Table 5.1 Classification of Metamorphic Rocks

TEXTURE	COMPOSITION			
	Aluminous (Shaly)	**Calcareous**	**Mafic**	**Quartzo-feldspathic**
Foliated	Slate Phyllite Schist	Calcareous schist	Greenschist Hornblende schist	
Foliated and layered	Gneiss	Calc-silicate gneiss	Mafic gneiss	Gneiss
Non-foliated	Granofels	Marble (calc-silicate granofels)	Greenstone Amphibolite Mafic granofels	Quartzite Quartzo-feldspathic granofels
Non-foliated	**Hornfels,** if definitely the result of contact metamorphism			

Metamorphic grade and grain size increase from top to bottom in foliated and non-foliated rows.

Table 5.2a (see Nature of the Parent Rock) lists key minerals that help to identify the parents of metamorphic rocks. Some minerals can form during metamorphism of more than one parent rock type. For example, biotite can form from both shales and basalts. In such cases, the complete assemblage of metamorphic minerals will help distinguish between the possibilities.

---

d) What were the parent rocks for the four specimens you studied?

Specimen __ **Specimen __** **Specimen __** **Specimen __**

Role of the Type of Metamorphism

Exercise 5.2 Metamorphic Textures

In Exercise 2.3d, you made a typical metamorphic rock with a **foliated** texture, a parallel alignment of platy "minerals."

What caused the "minerals" to become aligned?

Heat breaks bonds between atoms and ions, making the original minerals unstable, and permitting the freed particles to move to new crystal seeds. There are several ways in which heat can be applied to rocks, including the intrusion of magma, or by burial. As a rock is buried, overlying materials act as a blanket, preventing loss of internal Earth heat. This causes heat to accumulate within the rock.

Pressure compacts rocks, forcing grains, and even the atoms in grains, closer together. Added pressure may come simply from being buried deeper in the earth, resulting in **lithostatic pressure,** which is equal in all directions. It may also come from being squeezed, as between two colliding lithosphere plates. This causes **directed pressure,** which is stronger in some directions than in others. Figure 5.1 shows the effect that the two kinds of pressure would have on a spherical mineral grain.

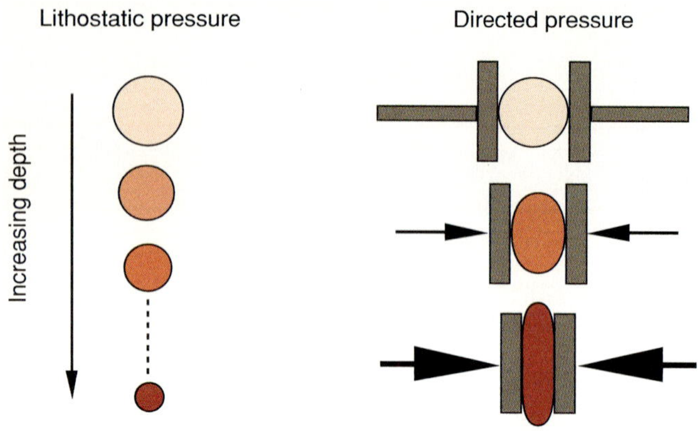

FIGURE 5.1 Effects of pressure on metamorphic rocks showing the effect on a spherical grain.

---

a) Which kind of pressure is involved in creating a foliation?

FIGURE 5.2 Metamorphic textures.

TABLE 5.2 INTERPRETING THE HISTORY OF METAMORPHIC ROCKS

A) NATURE OF THE PARENT ROCK

	Shale	Limestone or Dolostone	Mafic Igneous (basalt or gabbro)	Felsic Igneous or Sandstone
Key Minerals	Andalusite Sillimanite Kyanite Staurolite	Calcite Dolomite Olivine	Serpentine	
Key Mineral Assemblages	Muscovite + biotite ± chlorite, garnet	Calcite + talc Calcite + actinolite Calcite + epidote Calcite + garnet	Plagioclase + epidote ± chlorite, actinolite Plagioclase + hornblende ± garnet	Quartz + potassic feldspar + plagioclase feldspar

B) INTENSITY OF METAMORPHISM (INDEX MINERALS)

	Shale	Limestone or Dolostone	Mafic Igneous (basalt or gabbro)	Felsic Igneous or Sandstone
Low	Chlorite Biotite	Talc Actinolite Epidote Olivine	Epidote Chlorite Actinolite	Can't Easily Determine Intensity
Medium	Garnet Staurolite Andalusite	Garnet Pyroxene	Hornblende Garnet	
High	Kyanite Sillimanite	Wollastonite	Pyroxene	

C) TYPE OF METAMORPHISM (REGIONAL VS. CONTACT)

Texture	Regional	Contact	Can't Tell
Foliation	▼		
Lineation	▼		
Granoblastic (with micas)		▼	
Granoblastic without micas			▼
Layered but not foliated			▼
Layered and foliated	▼		

▼This property is rarely seen or measurable.

Special names are given to some kinds of foliation. When the minerals that are foliated are micas such as biotite and muscovite, the foliation is called **schistosity** (Figure 5.2a). Sometimes a type of layering forms during metamorphism in which layers with a foliated texture alternate with non-foliated layers. The combination of layering and foliation is called **gneissosity** (Figure 5.2b,c). **Lineation**, the parallel alignment of elongate minerals, forms when a rock containing rodlike, rather than platy, minerals is subjected to this kind of pressure.

Not all rocks develop the preferred orientation of mineral grains that characterizes foliated and lineated textures. Instead, some exhibit a random orientation of grains, regardless of grain shape, called a **granoblastic** texture (Figure 5.2e); a rock with a granoblastic texture is called a granofels.

b) Suggest two possible explanations for a rock to have a granoblastic texture instead of either a foliated or lineated texture.

Identifying the Type of Metamorphism The two principal types of metamorphism are called **contact metamorphism** and **regional metamorphism.** In contact metamorphism, heat is applied by the intrusion of magma; little if any pressure is involved. In regional metamorphism, lithosphere plates collide, adding directed pressure to the heat and lithostatic pressure already present deep in the crust.

c) Two rocks containing abundant muscovite, biotite, and quartz have very different textures: one is strongly foliated, the other granoblastic. Which type of metamorphism is responsible for each?

d) What can you say about the type of metamorphism involved in formation of a granoblastic rock composed of quartz and potassic feldspar? Explain.

Table 5.2c shows how the texture of a metamorphic rock can help determine which type of metamorphism has occurred.

Role of Metamorphic Intensity

Metamorphic rocks respond both mineralogically and texturally to increasing amounts of heat and pressure. Metamorphic minerals form as heat is applied, but later break down when more heat and pressure are applied than their structures can withstand. Thus, individual minerals may serve as indicators of the intensity of the metamorphic agents. These are called **index minerals** and are shown in table 5.2b.

Grain Size and Metamorphic Intensity Grain size also changes with increases in metamorphic intensity, primarily because of the increase in heat. In contrast to igneous rocks, grain size is not controlled by the rate of cooling, but rather by the amount of heat applied and the amount of time that the rock stayed hot.

e) Would you expect grains to be coarsest in a low-grade metamorphic rock (formed under conditions of low metamorphic intensity) or a high-grade metamorphic rock? Explain.

Non-uniform grain sizes in metamorphic rocks result not from different cooling rates, but from different numbers of seed crystals. Fewer seeds result in coarser grains. A rock with a few very large crystals is said to have a **porphyroblastic** (not porphyritic) texture (Figure 5.2d), and the large grains are called **porphyroblasts** (not phenocrysts).

Classification of Metamorphic Rocks

As with the other two kinds of rocks, our classification scheme is based on a combination of mineralogy and texture. Table 5.1 is a simplified classification scheme. First, identify the minerals present and determine the composition of the rock using table 5.2a. This places you in the appropriate column. The texture will control what row contains the rock name. In some cases, there will be only one choice, and the name is found. In others, there is more than one possibility; the following descriptions will help you make the correct identification:

Slate A slightly metamorphosed shale composed largely of foliated flakes of muscovite that are too small to be seen. The foliation is revealed by the tendency of slate to break into thin slabs parallel to the foliation. Colors range widely, and include gray, black, red, and green (Figure 5.3a).

Phyllite A more intensely metamorphosed shale (or a metamorphosed slate) composed largely of foliated flakes of muscovite ± chlorite ± biotite. These grains are somewhat larger, but individual flakes may still be difficult to discern. The foliation is revealed by the shiny surface of the rock, caused by the reflection of light from the parallel mica flakes (Figure 5.3b).

Schist A still more intensely metamorphosed foliated rock with the composition of shale, in which individual mica flakes are easily identified. Porphyroblasts of aluminum-rich minerals such as garnet may be present (Figure 5.3c).

Gneiss A rock composed of light and dark layers, one of which is foliated.

Greenschist A slightly metamorphosed mafic rock composed of chlorite and plagioclase feldspar, with epidote and/or actinolite. Foliation is defined by parallel alignment of chlorite. These are truly green schists, because chlorite, epidote, and actinolite are all various shades of green (Figure 5.3d).

Hornblende Schist A higher grade version of greenschist, in which foliated (and often lineated) hornblende crystals have replaced the chlorite, epidote, and actinolite.

Greenstone A nonfoliated rock containing the same minerals as a greenschist (Figure 5.3e).

Amphibolite Similarly, a nonfoliated rock comparable in mineralogy to hornblende schist.

Quartzite A monomineralic, nonfoliated (granoblastic) rock composed almost entirely of quartz (Figure 5.3f).

Marble A granoblastic rock composed mostly of calcite or dolomite.

Hornfels *Any* rock that is formed as the result of contact metamorphism may be called a hornfels, but you must be absolutely certain that this is the case. If not, the term granofels is preferred, since it carries no genetic connotation. Typical hornfelses are extremely fine-grained, dense granofelses.

Exercise 5.3 Identification of Metamorphic Rocks and Interpretation of Metamorphic History

Your ultimate goal in analyzing a metamorphic rock is to determine its complete history, including the nature of the parent rock, the type of metamorphism involved, and the intensity of metamorphism. Collect your data in the sheets provided at the end of this chapter and use tables 5.1 and 5.2 to find the name of each rock in your set of specimens and deduce its history.

FIGURE 5.3 Examples of some common metamorphic rocks.

DATA SHEET FOR METAMORPHIC ROCKS

NAME ____________________

SPECIMEN NUMBER	MINERALS PRESENT (MOST ABUNDANT FIRST)	TEXTURAL FEATURES	NAME OF ROCK	ORIGIN OF ROCK

Earth's History Determined from Its Surface Features

Section 2

In Section 1, you learned how to interpret Earth processes and history from minerals and rocks. You can now identify different rock-forming processes from the rocks they produce and use textures to infer the rate and location (intrusive or extrusive) of magmatic cooling, the kinetic energy of the agent responsible for transporting particles found in a clastic sedimentary rock, and the kind of metamorphism that affected an area. You can also use your knowledge of minerals to estimate metamorphic conditions or ancient climates.

In this section, you will learn how to use a completely different set of geologic clues to interpret the past and present behavior of the Earth. Major landscape features, such as mountain ranges, stream valleys and canyons, and the shapes of shorelines, reveal the processes involved in the development of the Earth's surface—to the educated eye. Chapters 6 through 11 will begin the necessary education, showing you how to "read" the information present in landscape features, many of which are probably already familiar to you.

Actually, the landscape in any area is the result of two enormously powerful and conflicting sets of processes. **Constructional** processes, those tectonic processes that raise mountain ranges during plate collisions or volcanic eruptions, are pitted against the **destructional** processes of erosion that attempt to level highlands and fill in lowlands. Interpretation of constructional processes is the topic of Section 3. Here, the focus will be on the agents of erosion that carved deep valleys, such as the Grand Canyon of the Colorado River in Arizona, or broad valleys, such as those of the Missouri, Mississippi, and Ohio in the midcontinent; deposited extensive blankets of bouldery and sandy sediment across the northeastern and north-central parts of our country; and formed the exotic shapes found in Bryce Canyon and other parks in the arid Southwest.

One of the most valuable tools in studying the surface of the Earth is a good map. Section 2 therefore begins with an introduction to mapmaking and mapreading, concentrating on the topographic maps that depict the landscape in great detail (Chapter 6). Once you are comfortable with the use of these maps, you are ready to analyze the destructional processes. Chapters 7 through 11 deal with the agents of erosion—streams, glaciers, groundwater, wind, and shoreline currents—and show how to recognize which has been active in a given area.

Sections 1 and 2 are closely connected to each other. You will quickly realize that erosion is just the removal of clastic and chemical particles from their source regions and that depositional features such as river deltas are simply sediments waiting to be lithified.

CHAPTER 6

INTRODUCTION TO MAPS

PURPOSE

To become acquainted with the elements of maps and to develop skills in reading topographic maps.

MATERIALS NEEDED

1) A clear plastic ruler (preferably one marked in tenths of an inch)
2) A protractor (circular, if you can get one)
3) Topographic maps (provided by your instructor)
4) A globe or world map

INTRODUCTION

A map is a representation of a portion of the Earth's surface that accurately portrays the locations, shapes, and sizes of the features that it is designed to show. Maps can be used for many different purposes, and are designed accordingly. Road maps show the highways that connect population centers, but do not usually depict topographic features very accurately. Weather maps show temperatures and different air masses, but do not deal with topography or highways. The type of map that we will be using is called a **topographic map,** because it is designed to show the topography of an area: the three-dimensional aspects of the landforms present, including size, shape, and height.

Every map must contain certain elements regardless of what it is intended to illustrate. These elements are **location,** some method for measuring **distance,** and careful attention to showing the correct **direction** between points on the ground. We will first look in some detail at different methods for showing these three elements, and then see how changes in elevation are shown on topographic maps.

Before getting down to business, however, a word of caution. The Earth is three-dimensional (a sphere), but most maps are flat, only two-dimensional. The process of taking a three-dimensional object and portraying it in two dimensions is called making a projection. There are several different kinds of projections, each with its own benefits and drawbacks. Each shows one particular map element precisely and distorts the others somewhat. Have you ever wondered why Greenland seems to be as large as North America on some maps, but not on others? Why Antarctica seems to be "wider" than all the rest of the continents? Figure 6.1 shows how a human head would appear if subjected to some of the common map projection methods. Most of the maps that you will be using in your laboratory cover only a small area of the Earth's surface, and are thus not badly affected by different projection methods. Note that the small, centrally located eyes and ears in figure 6.1 are relatively undeformed compared to the larger-scale shape of the skull. Distortion is greatest for areas near the North and South poles; very different methods must be used to accurately represent the polar regions.

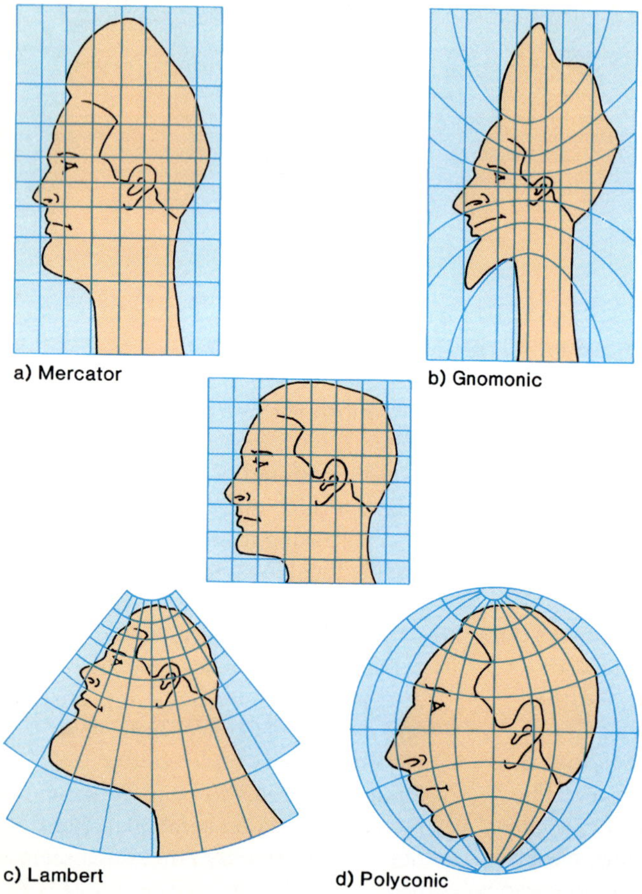

FIGURE 6.1 A human head seen in common map projections.

MAP ELEMENTS

Location

A good map must accurately show the locations of the features present in the area that it covers, and ideally should locate that area in the larger picture—on the Earth as a whole. Most road maps use a simple grid system involving letters and numbers to locate cities and towns. Several different grids have been developed that are extremely accurate, and we shall look at those that appear on the maps you will be using later in this course.

Latitude and Longitude The latitude/longitude system for locating points on the Earth's surface has been successfully used for hundreds of years. Figure 6.2 shows how this grid covering the entire world is created. A value of latitude tells how far north or south of the equator a point is located, in degrees of arc. As shown in figure 6.2a, all points located 30° north of the equator lie along a single line (called a **parallel of latitude** because it is parallel to the equator). Points 60° north, 30° south, or 27°37′42″ (read 27 degrees, 37 minutes, 42 seconds) south of the equator would lie on different parallels. Remember that there are 60 minutes in one degree and 60 seconds in one minute of arc.

The equator is an obvious baseline for latitude, but there is no comparable reference line for longitude. Imagine the Earth cut vertically into sections like the sections of an orange or grapefruit. Select one of the cuts as a reference plane, and measure the angular distance east or west of that plane (Figure 6.2b). The arbitrarily selected reference line is called the **Prime Meridian** and passes through the Royal Observatory at Greenwich, England; it continues on the opposite side of the Earth as the International Date Line. All points that are the same angular distance from the Prime Meridian are said to lie on a **meridian of longitude.** Note that, unlike parallels of latitude, meridians of longitude approach one another and meet at the poles.

To successfully locate any point on the surface of the Earth, all that is needed is a set of latitude and longitude coordinates. For example, 45°22′38″N, 67°18′58″W accurately locates a point in eastern Maine. Latitude is always given first, but there should be no confusion about that since only latitude can have the North notation. Note that simply writing 45°22′38″ is not good enough. Even if you knew that this is a latitude value, you would not know whether it was in the Northern or Southern Hemisphere. Similarly, there must be an indication of whether the longitude value is east or west of the Prime Meridian. The maximum value of latitude is 90°. Ninety degrees north latitude is the North Pole, 90° south latitude the South Pole. By convention, the maximum longitude value is 180° east or west (the International Date Line).

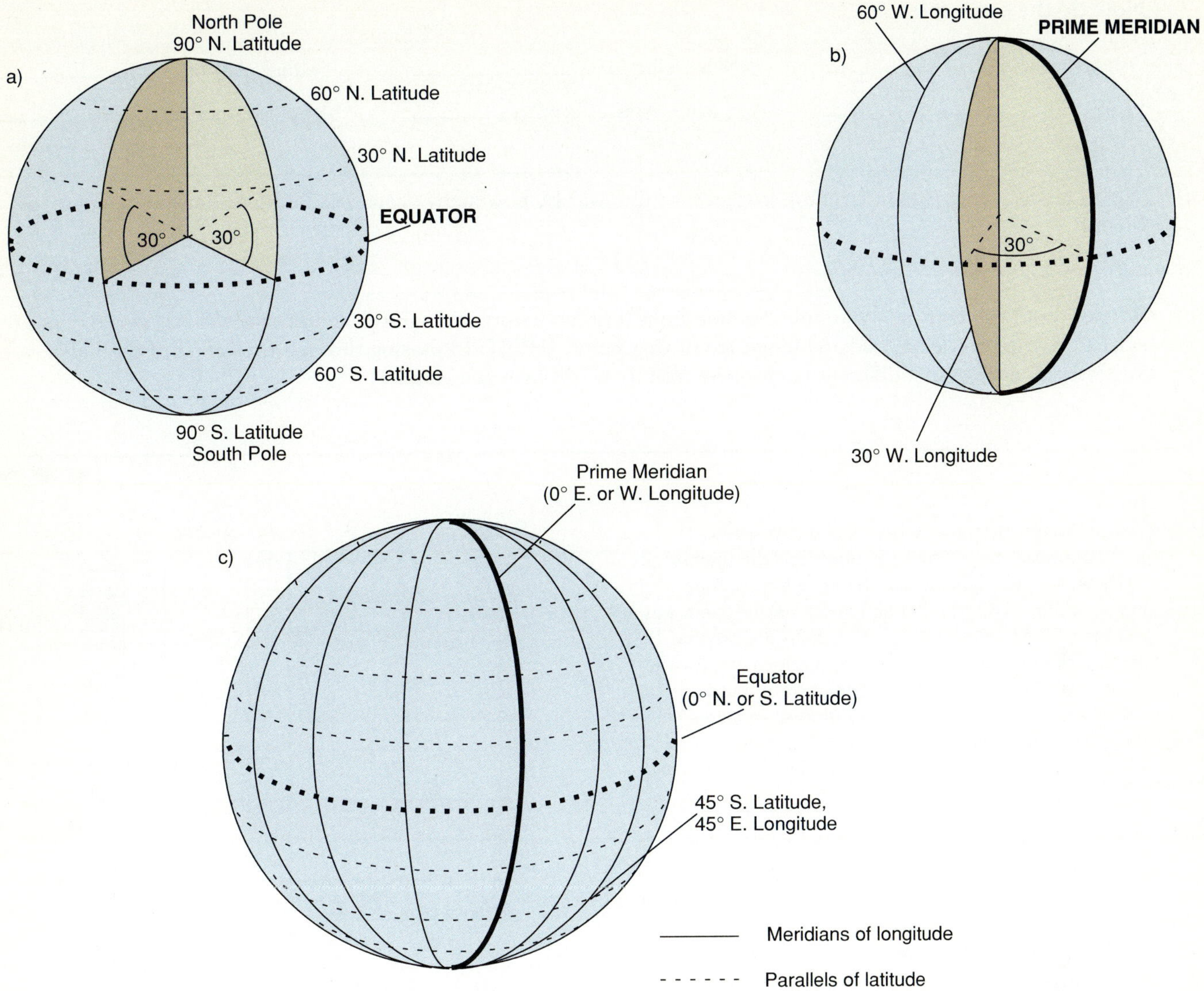

FIGURE 6.2 Construction of the latitude-longitude grid. a) Each parallel of latitude contains all points located at a specific angular distance north or south of the equator. b) Each meridian of longitude contains all points located at a specific angular distance east or west of the prime meridian. c) The latitude-longitude grid.

EXERCISE 6.1 USE OF LATITUDE AND LONGITUDE FOR LOCATION

a) With the aid of a globe or world map, determine the latitude and longitude of your geology laboratory as accurately as you can.

____

b) What are the latitude and longitude coordinates of your home town?

____

c) Determine the latitudes and longitudes of the following cities in North America:

Mexico City, Mexico ____________ Seattle, Washington ____________ Miami, Florida ____________

Nome, Alaska ____________ Quebec City, Quebec ____________ Los Angeles, California ____________

St. Louis, Missouri ____________ New York, New York ____________ Ottawa, Ontario ____________

Now do the same for:

Oslo, Norway ______ New Delhi, India ______ Sydney, Australia ______

Beijing, China ______ Buenos Aires, Argentina ______ Baghdad, Iraq ______

______

d) Which city is farther north: London, England, or Ottawa? By how many degrees of latitude? Barcelona, Spain, or Miami?

______

e) Where would you emerge if you drilled a hole from your laboratory through the center of the Earth to the opposite side? Give the precise latitude and longitude of that point. (HINT: In passing through the center of the Earth, you are going to emerge in a different hemisphere from that in which you started.)

Public Land Survey System When Americans spread westward from the original 13 states into the uncharted "wild west," they quickly saw the need for accurate maps, and in 1785, the Public Land Survey System was initiated. Today, most states, other than the original 13, are covered by this system. (The exceptions are, for the most part, those settled and surveyed by Spanish colonists: Texas and the other southwestern states.) It provides a grid capable of easily locating points to within an eighth of a mile with just a few coordinates.

Figure 6.3 shows how this grid is generated. Two lines are surveyed with extreme accuracy in a given region, one running north-south (the **Principal Meridian,** not to be confused with the *Prime Meridian*), and one running east-west (the **Base Line**). Secondary lines are surveyed at distances 6 miles apart that are parallel to the Principal Meridian and the Base Line. These effectively create a grid composed of squares 6 miles on a side (Figure 6.3a). These squares define a series of east-west rows called **Townships** and north-south columns called **Ranges.** All points in the "row" up to 6 miles north of the Base Line are said to lie in Township 1 North (T1N); those lying between 6 miles south and 12 miles south of the Base Line would be in T2S. Similarly, all points in the "column" between 12 miles east and 18 miles east of the Principal Meridian would be in Range 3 East (R3E).

Thus, the coordinates T2S, R5E locate a point to within the 36-square-mile area shown in heavy shading in Figure 6.3a. This square is then divided into 36 squares called **sections,** each one mile on a side and numbered as shown in Figure 6.3b. The designation Sec. 12, T2S, R5E locates a point to within one square mile. For even greater accuracy, sections are divided into quarters (Figure 6.3c), and each quarter-section can be further divided into fourths (Figure 6.3d). The system permits location to within ¼ mile and, with a simple estimate, to within ⅛ mile.

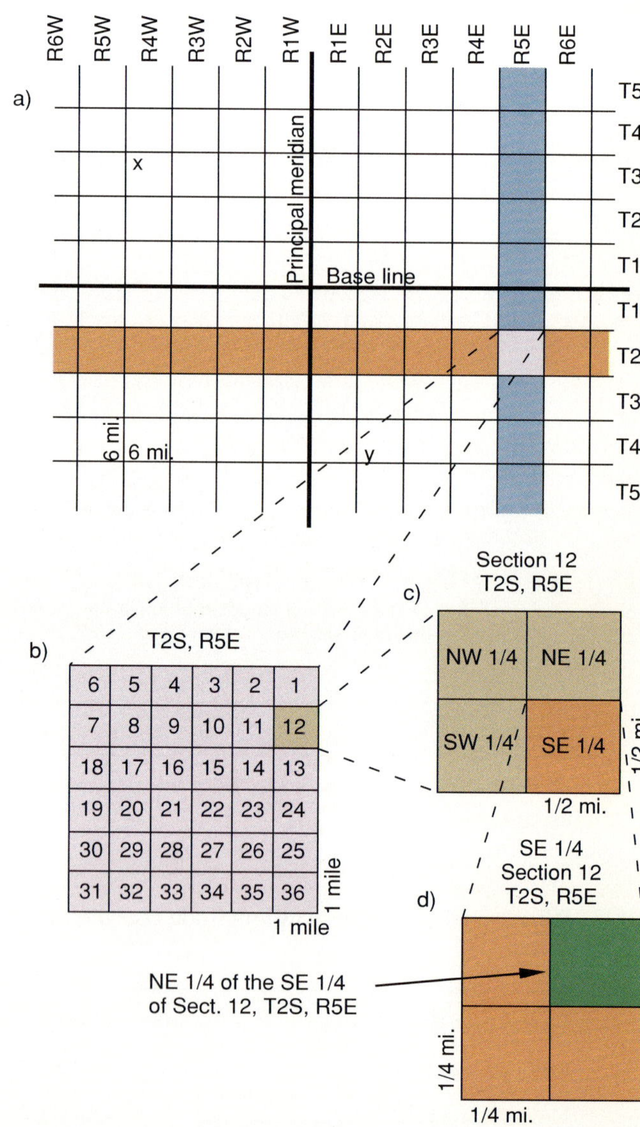

FIGURE 6.3 Construction of the township-range grid. a) East-west lines spaced six miles apart define townships, north-south lines spaced six miles apart define ranges. Their intersections create squares 6 miles on a side identified by the particular township and range that they belong to. b) Each such square is divided into 36 sections, numbered as shown. c) Each section is divided into quarters. d) Each quarter-section is divided into fourths.

EXERCISE 6.2 A LITTLE PRACTICE WITH THE PUBLIC LAND SURVEY SYSTEM

a) Give the location as accurately as you can of the points labeled X and Y in figure 6.3.

X ____________________

Y ____________________

---

b) Give the coordinates of points located:

1) 14 miles north of the Base Line and 34 miles west of the Principal Meridian ____________________

2) 44 miles south of the Base Line and 50 miles east of the Principal Meridian ____________________

Direction

A good map must accurately show the precise direction between any two points on the Earth's surface, so that you could follow that direction with a compass and arrive at the desired location. Directions are usually measured with respect to geographic north—the direction from a starting point toward the North Pole. A compass needle points to **magnetic north,** which is a point defined by the Earth's magnetic field and does not coincide with the geographic (true) north pole (Figure 6.4a). The difference between true north and magnetic north is called magnetic declination, and must be taken into account when using a compass. Declination has different values throughout the world; at various locations in North America, compass needles point west of true north (west declination) or east of true north (east declination), as shown in figure 6.4b.

For example, consider points A and B in figure 6.4b. At A, a compass needle would point about 15° east of true north (declination is 15°E), whereas at B, it would point 15° west of true north (declination at B is 15°W). Only a few lucky people live in an area where a compass needle actually points to true north.

The **azimuth method** is a simple way to indicate direction, based on the dial of a compass (Figure 6.5a). If you face north and turn slowly clockwise, you will face east after turning 90°, south after 180°, and west after turning 270°. After a full 360°, you are facing north again. These angular relations provide a highly accurate way of describing direction. The compass direction 180° uniquely and unmistakably means south, and 270° means west. Similarly, 220° must be halfway between due south and due west (southwest).

Consider figure 6.5b. To determine the precise direction from Point A to Point B, simply set up your compass over Point A and draw a line to B. This line intersects the compass dial at about 100°, so the direction (or **bearing**) of Point B from Point A is said to be 100°. Note that the bearing of Point A from Point B is exactly the opposite of this bearing (280°). In the same way, the bearing of Point C from Point B is 220°. By convention, all bearings are three-digit numbers. Thus, a bearing of due east would be written as 090°; northeast would be 045°.

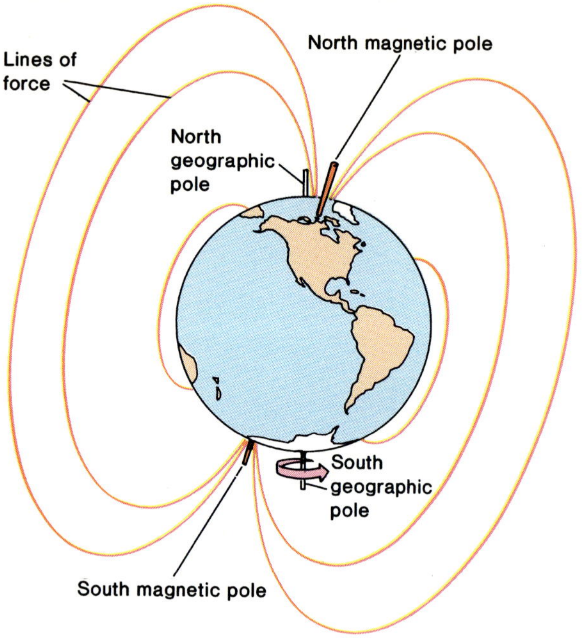

a)

FIGURE 6.4 a) Magnetic poles and magnetic declination showing the relationship between magnetic and geographic (true) north. b) Magnetic declination in North America. (Reprinted with the permission of Merrill an imprint of Macmillan Publishing Company from *Physical Geology* by Coch and Ludman. Copyright © 1991 by Macmillan Publishing Company.)

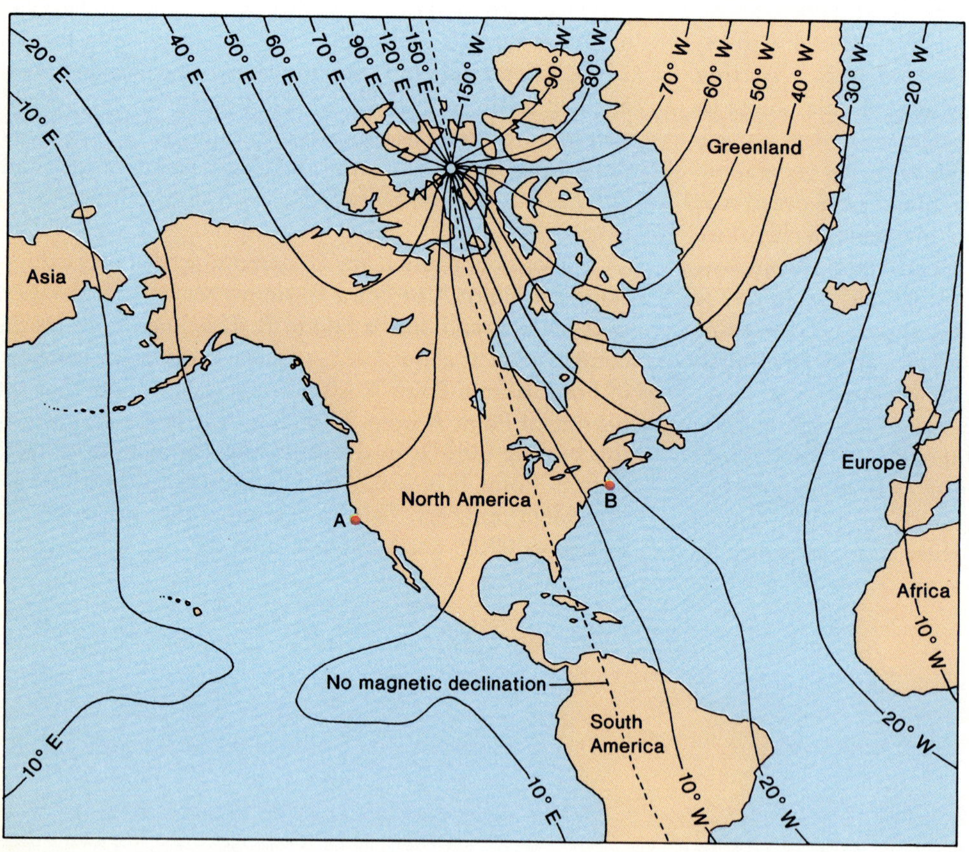

b)

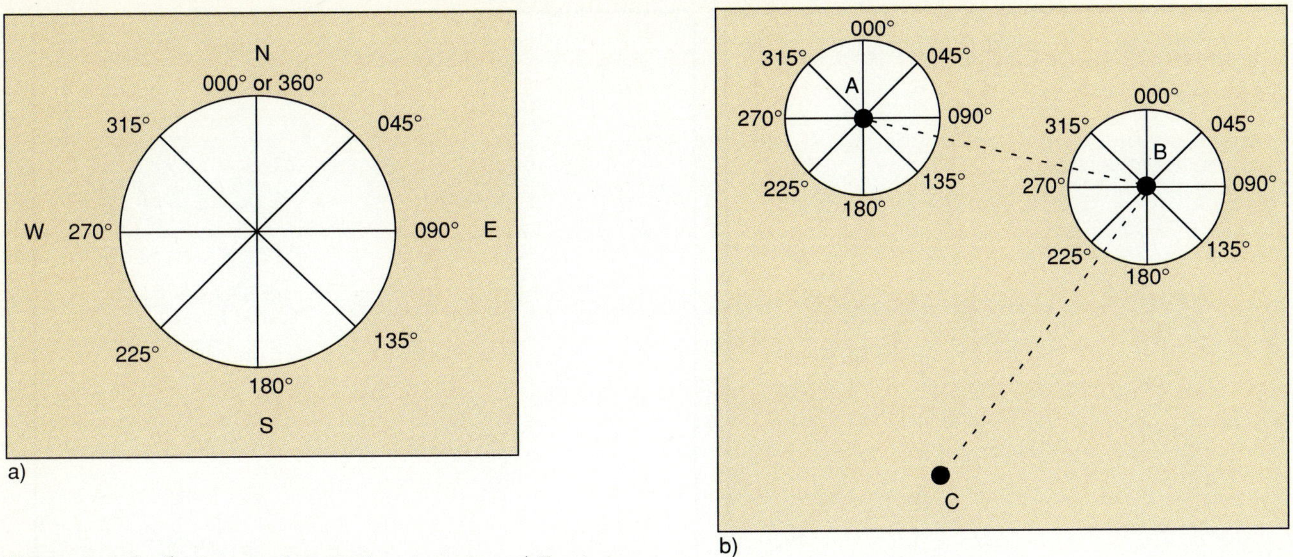

FIGURE 6.5 Compass bearings between points. a) Bearings are given as 3-digit angular values, e.g., Northeast=045 degrees, Southwest=225 degrees. b) The bearing from A to B is 100 degrees, from B to A, the bearing is 280 degrees.

EXERCISE 6.3 HOW NOT TO GET LOST

a) In figure 6.5, what is the bearing of point B from point C?

b) What is the bearing of point A from point C? Of point C from point A?

c) Return to figure 6.3. What is the bearing of point X from point Y? point Y from point X?

Distance

We can fit a map of anything onto a sheet of 8½ × 11″ paper; all we have to do is scale down the area so that it fits. Of course, to fit the page, we would have to scale down the entire world much more than would be needed for one county in Montana. We would lose a lot of detail in fitting the world on a page, but it could be done. Figure 6.6 shows a single area mapped at three different scales. Why use different scales? We select the scale that shows most clearly what we want our map to depict. Thus, figure 6.6a would be appropriate if we wanted to show the locations of several towns; figure 6.6b for more detail of a smaller area; and figure 6.6c for a closeup view of individual streets, driveways, and houses.

Each map must tell exactly how much the actual area has been scaled down; only then can you measure the distance between two cities on the map and know how far apart they are in the real world. That way, you can plan driving times or estimate how far you will have to hike through the woods to get somewhere. There are three ways to tell the scale of a map: verbally, proportionally, and graphically.

A **verbal scale** uses words to describe the relation between dimensions on the map and those on the ground. For example: "one inch equals approximately 7.5 miles" is a common verbal scale found on roadmaps. It is easy to use but is not very accurate.

A **graphical** or **bar scale** (Figure 6.7) uses a line on the bottom of the map to express the same relation as the verbal scale, dividing it into inches and fractions of an inch and marking the appropriate true ground distance above the divisions. This can be more accurate than a verbal scale, depending on how carefully you measure distances.

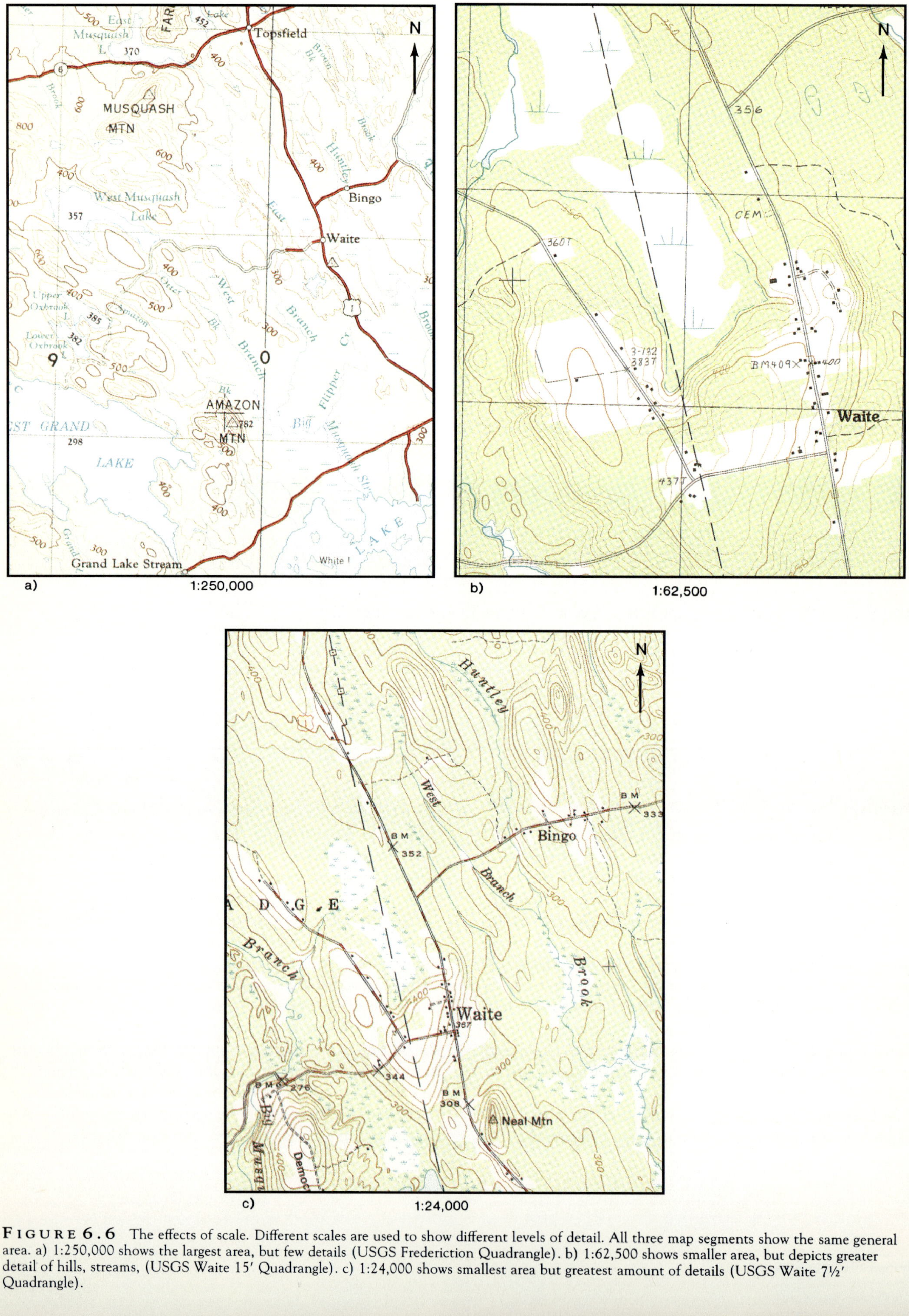

FIGURE 6.6 The effects of scale. Different scales are used to show different levels of detail. All three map segments show the same general area. a) 1:250,000 shows the largest area, but few details (USGS Frederiction Quadrangle). b) 1:62,500 shows smaller area, but depicts greater detail of hills, streams, (USGS Waite 15′ Quadrangle). c) 1:24,000 shows smallest area but greatest amount of details (USGS Waite 7½′ Quadrangle).

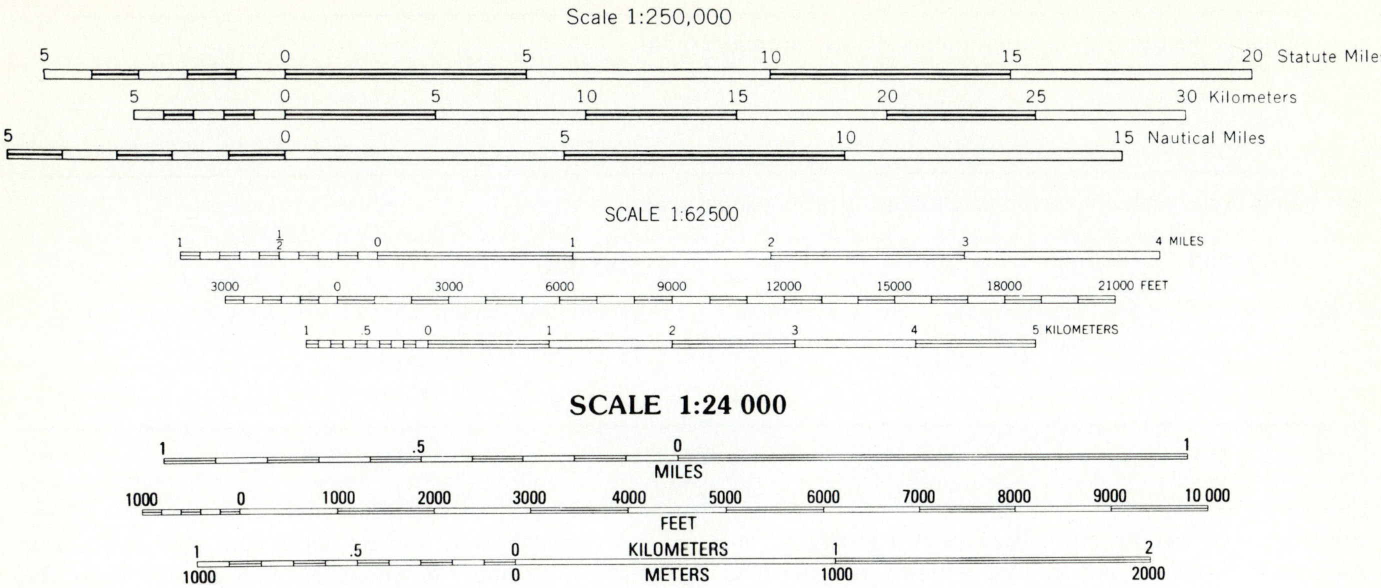

FIGURE 6.7 Bar and proportional scales.

The most accurate way of indicating scale is to use a **proportional scale,** one that precisely indicates how much the ground has been shrunk to fit on the paper. Proportional scales look like this: 1:100,000 (read: one to one hundred thousand), meaning that one unit of length on the map is equivalent to 100,000 of the same units on the ground. On such a map, all distances on the ground have been scaled down to 1/100,000 of their true size. Another way to look at it is to say that a map distance must be multiplied 100,000 times to equal the true distance on the ground.

What unit of length does the scale refer to? All units! One inch on the map equals 100,000 inches, one foot on the map equals 100,000 feet, and 1 cm on the map equals 100,000 cm on the ground. Because it uses units that are 10, 100, or 1000 times greater than others, the metric system is ideally suited for scales such as 1:100,000 or 1:50,000. The relation between inches, feet, and miles is not nearly as simple, so that maps designed for use in the United States commonly have scales that look more complex. For example, distances on a map are most easily measured in inches, but ground distance is measured in miles. There are 63,360 inches in a mile (12 inches/foot × 5,280 feet/mile), so that the proportional scale 1:63,360 means that one inch on the map corresponds exactly to one mile on the ground. Many maps use a scale that is close to this, 1:62,500; this is only about 70 feet short of a mile, and for most of our purposes in the lab, can be considered to be equivalent to one inch representing a mile.

EXERCISE 6.4 USING SCALE TO MEASURE DISTANCE ON A MAP

a) Fill in the appropriate proportional scales:

Verbal Scale	**Proportional Scale**
One inch on the map = 1 mile on the ground	____________
One inch on the map = 2,000 feet	____________
One inch on the map = 4 miles on the ground	____________
One inch on the map = 8 miles on the ground	____________

b) What is the scale used in figure 6.3a? (How did you figure it out?)

c) What is the distance between points X and Y in figure 6.3a?

d) Draw a bar scale for each of the following proportional scales:

1:62,500 **1:500,000**

EXERCISE 6.5 PUTTING THE MAP ELEMENTS TOGETHER AND USING THEM

While on your way to join colleagues at a geological research station on the remote island shown in figure 6.8, your plane crashes on the shore of the island. You and two colleagues are left at the crash site without a radio or other means of communication. Your sketch map shows that the island is dangerous, with many hazards between you and the research station. As the only member of your group who knows how to use a compass, you are the one who must navigate a route to safety.

Unfortunately, you are not in very good shape; none of you can climb a mountain or swim across a river. The inhabitants of the island (other than the geologists) are rather unfriendly. It would be best if you could come no closer than 1.5 miles to any part of their territories. You can carry enough water for 3 days, and can walk about 12 miles per day.

a) Sketch the most direct feasible route to the research center.

b) Will you have to stop for water? On which day?

c) List the directions and distances that you must follow for each leg of the journey in order to reach safety.

TOPOGRAPHIC MAPS

The most useful maps for studying landforms are topographic maps. In order to show topographic features—hills, valleys, mountains, cliffs—these maps must depict an additional element: **elevation.** To do so, cartographers use lines called **contour lines** to show how elevation varies within the area covered. Examine figure 6.9, a topographic map of the island sketched in figure 6.8. The contour lines are brown, as they are on most topographic maps. Each line connects points that are the same elevation above sea level. Note that every fifth line is slightly darker than its neighbors, and is numbered; the number gives the elevation in

d) How far will you have to walk, and how long will it take?

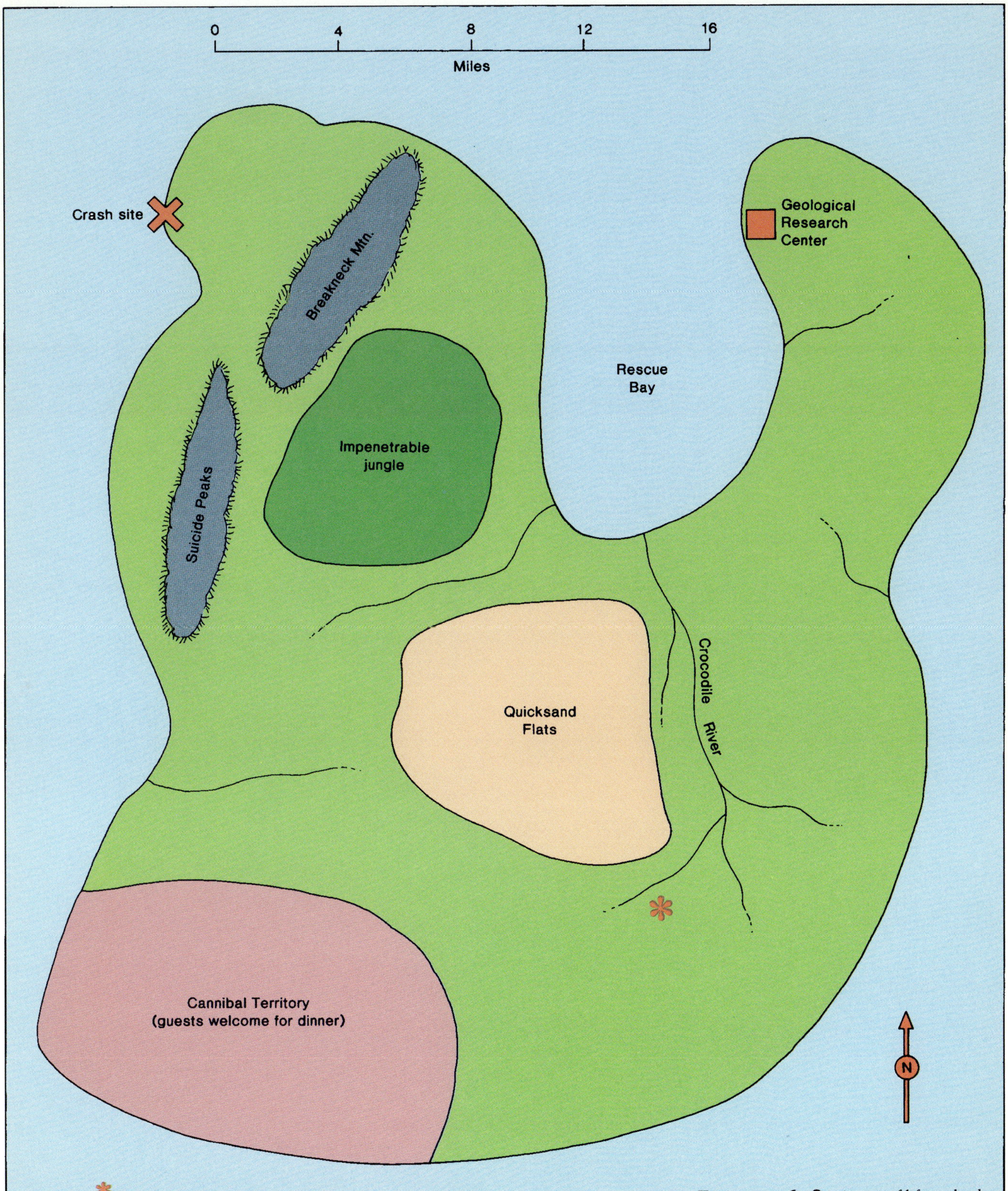

FIGURE 6.8 A case of life or death.

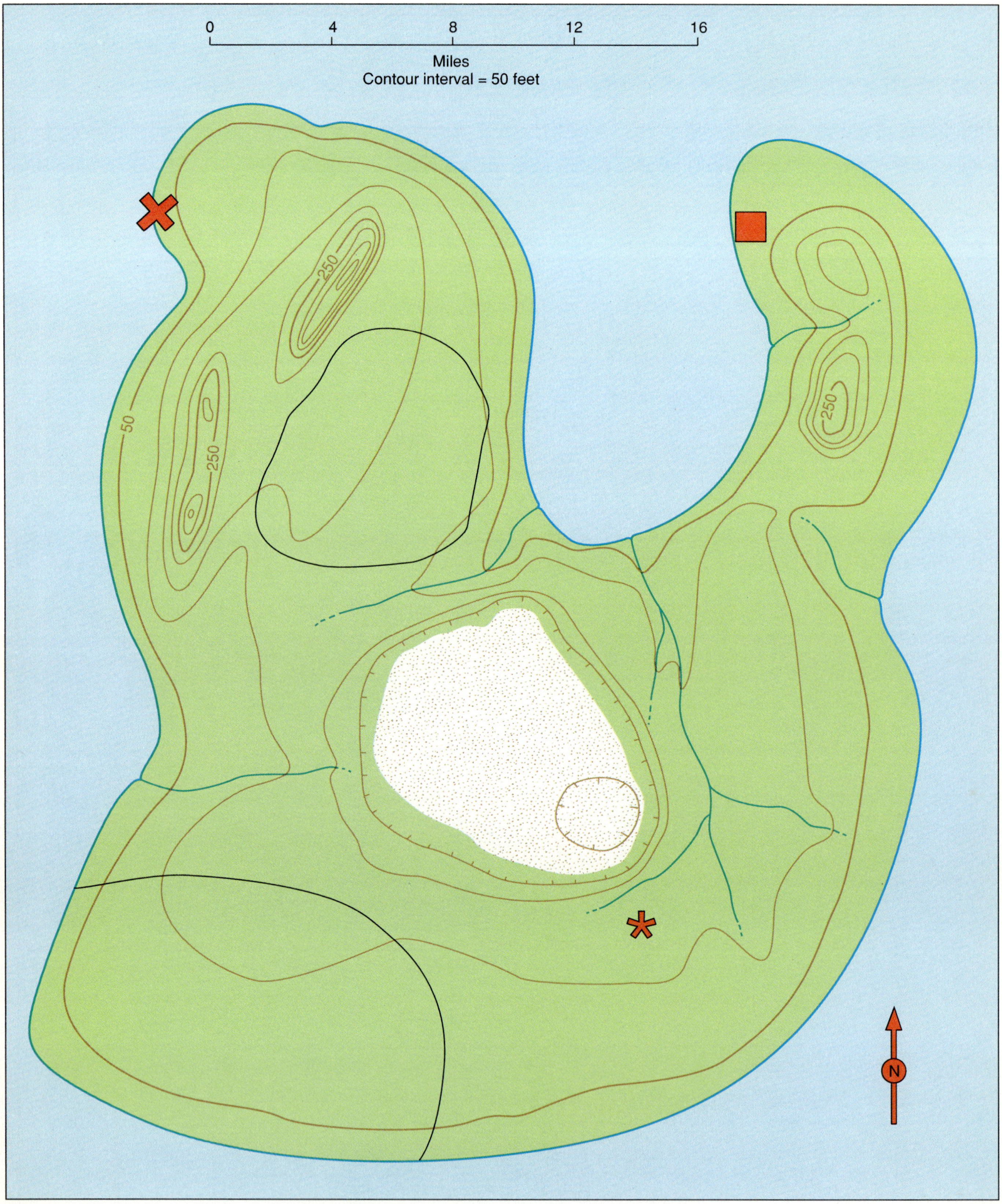

FIGURE 6.9 Topographic map showing details of the landforms on the island shown in figure 6.8.

feet of every point on that line. All points on one side of a contour line must be at a higher elevation than that line; all points on the other side must lie at lower elevations.

To determine the elevation on any line that is not numbered, look at the bottom of the map for the **contour interval.** This tells you the elevation spacing between lines. For this map, the contour interval (C.I.) is 50 feet, meaning that contour lines will have values that are multiples of 50; for example 100′, 150′, 200′, 250′, etc. Small contour intervals (10 or 20 feet) are used to show details in areas where there is not much elevation change. Larger intervals (100, 250 feet) are used in mountainous regions where there are rapid elevation changes.

Reading a contour map is a skill that can be acquired with only a little practice. Just keep the definition of a contour line in mind, and remember that points on opposite sides of a contour line will be higher or lower than the line itself. Common sense will provide all the additional "rules" that are needed.

EXERCISE 6.6 ESTABLISHING THE "RULES" OF CONTOUR LINES

a) Determining elevations

Why, in normal circumstances, is it impossible for two contour lines to cross each other? (HINT: Remember the definition of a contour line.)

Not every point on the island lies on one of the contour lines. How can you estimate the elevation of a point that lies between two contours?

What does a series of concentric contour lines indicate (as in the area just southeast of the research center)?

Estimate the elevation of the highest point on the Suicide Peaks. What limits are there on the maximum height?

Where is the highest point on the island? What is its elevation?

Locate and give the elevation of the highest point in the eastern part of the island.

In most instances, adjacent contour lines indicate different elevations. There are three common exceptions to this rule, situations in which adjacent contour lines have exactly the same elevation. All occur when there is a change in slope direction (up vs. down), as shown in figure 6.10: a) crossing a river valley (you first walk down into the valley, and then up the other side); b) climbing over the top of a hill or ridge (you first go up, then down the other side); and c) entering a depression (a variation on the stream-valley type)

Locate examples of each of these situations on figure 6.9.

---

b) Depressions

Contour lines with small tick marks (called hachures) indicate that elevation is **decreasing** on the side with the hachures. The contour interval is the same as for contour lines without hachures.

What is the elevation of the hachured contour line that surrounds Quicksand Flats?

What is the lowest elevation in Quicksand Flats?

---

c) Steepness of slope

Some parts of the map contain many contour lines; others few. What does this reveal about the differences in the land surface in those areas?

Along Breakneck Mountain, the contour lines are very close together, whereas in the Cannibal Territory, they are widely spaced. What does this tell you about the difference in the slope of the land surface in these areas?

Which side of Breakneck Mountain (NW or SE) is steeper? How do you know?

---

d) Stream valleys

Which way is the Crocodile River flowing? (HINTS: Water always flows downhill; the elevation of the shoreline is 0 feet above sea level.)

Follow the 100-foot contour line from the northeast as it crosses the Crocodile River. It bends sharply as it does so, and bends again on the opposite side, forming a "V." Note that the 50- and 150-foot contour lines do the same. Now look at the other streams. From the V of the contour lines alone, which way do these streams flow? (Check with the elevations to make sure your conclusions are correct.)

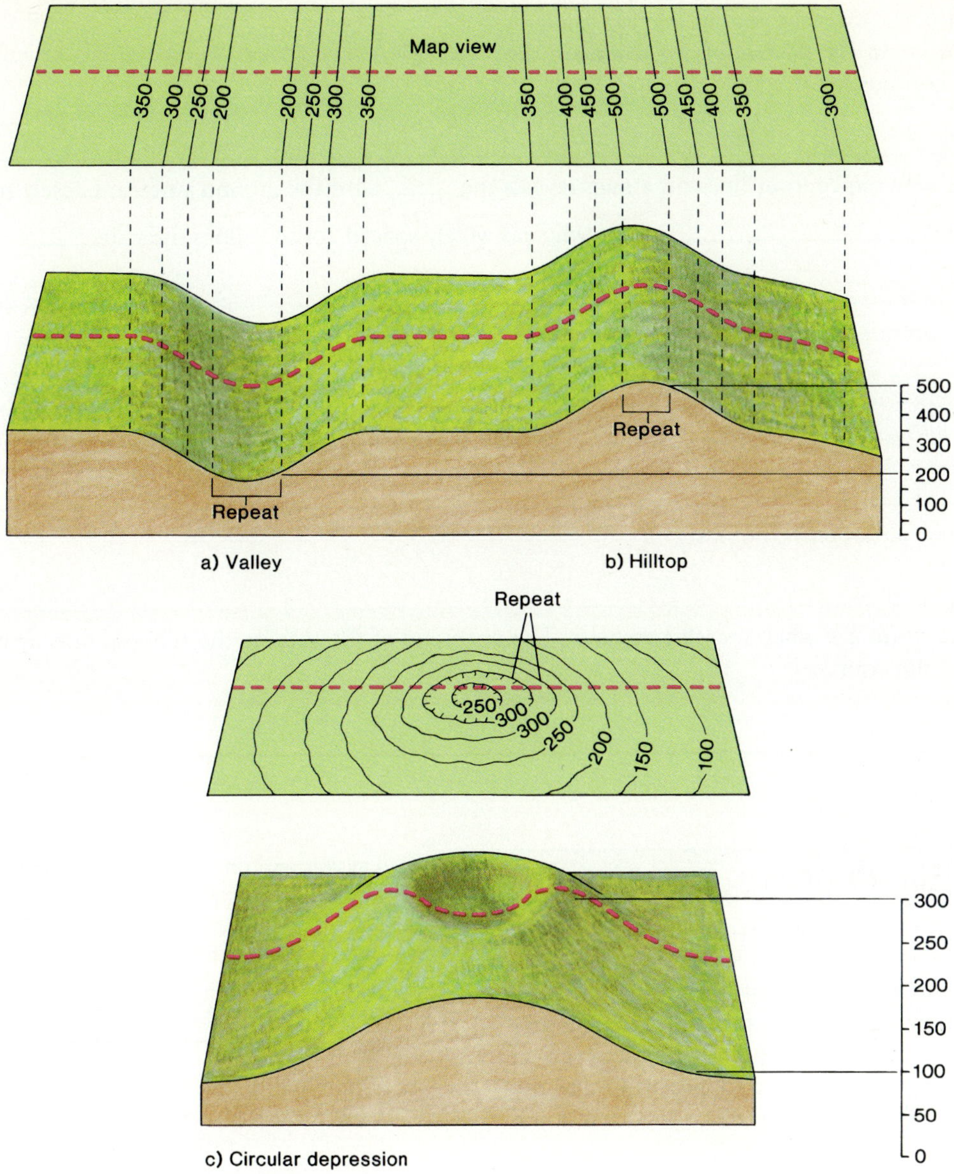

FIGURE 6.10 Elevations along adjacent contour lines.

Notice that the topographic map is much more useful than the simple sketch map, because it not only shows where features are, but also indicates their exact heights, steepness of hillslopes, and depth of river valleys.

Describe your route from the crash site to the research center, carefully indicating the nature of the uphill and downhill legs of your journey. How would you have to replan your route in view of information from the topographic map?

Congratulations! Now that you are resting safely at the research station, you should be able to complete the following rules of contour lines:

---

a) The spacing between contour lines on a map reveals the ________ of the ground surface. Closely spaced contour lines indicate ______________________, whereas widely spaced contour lines indicate ____________________.

---

b) Concentric contour lines indicate the presence of a ____________.

---

c) Concentric *hachured* contour lines indicate the presence of a ____________.

---

d) Contour lines cannot cross, because:

---

e) Contour lines form a V when crossing streams. The open end of the V faces the (choose one: upstream or downstream) direction.

Topographic Profiles

Although a contour map portrays the configuration of the Earth's surface in all three dimensions, it is still a flat, two-dimensional diagram. With the detailed information that a map provides, it should be possible for you to draw a **topographic profile:** a cross-section that shows the shape of the land surface along any given path. Figure 6.11 and the following steps will show you how to construct a topographic profile:

1) Select the line along which you wish to draw the profile (Line A-A′ in figure 6.11).
2) Study the line of the profile to determine the highest and lowest points that will be crossed.
3) On a separate sheet of paper, draw a line equal in length to A-A′ and label the ends A and A′. Sketch your concept of how the profile will look (this is a safety device: if you think there will be a deep valley in one place on the profile, and the completed profile shows something similar to Mt. Everest, you have either drawn the profile incorrectly or read the map wrong earlier).
4) On another piece of paper, draw the line A-A′. You should use graph paper for this part of the process. If not: beneath the first line, draw several parallel lines of the same length and label them with the same contour spacing that is present on the map. Select an elevation for the top line that is higher than the highest point of the profile and one for the bottom line that is lower than the lowest point on the profile.
5) Lay this grid on the map along the line of profile. Start at one end of the profile; wherever the line of profile crosses a contour line, draw a line perpendicular to your grid lines to the appropriate elevation, as shown in figure 6.11.
6) In a similar way, locate on your grid the elevations of the streams, hilltops, houses, roads, and other features that your profile crosses.
7) Connect the dots as shown to construct the profile.

Now compare that profile with the sketch that you originally drew. Ideally, the two should be the same, but chances are that your profile will make small hills look like the Alps, and small valleys look like the Grand Canyon. (Look, for example at Breakneck Mountain on the completed profile in Figure 6.11.) The reason for this is a problem of scale. If you had used exactly the same scale in your grid that was shown for the map, your hills barely would have looked like hills—they would have looked like tiny wiggles in a straight line! If one inch on the vertical profile represents less distance than one inch horizontally on the map, it will take more inches to show an elevation change of 1,000 feet than to show a map distance of 1,000 feet. This is how small hills become Alps.

The phenomenon described above is called **vertical exaggeration.** Profiles generally have at least some vertical exaggeration built in, because most changes in elevation are small compared with distances between points on the surface. At a scale of 1 inch = 1 mile (1:63,360), a 1,000-foot cliff would only require about 0.2 inches in a profile,

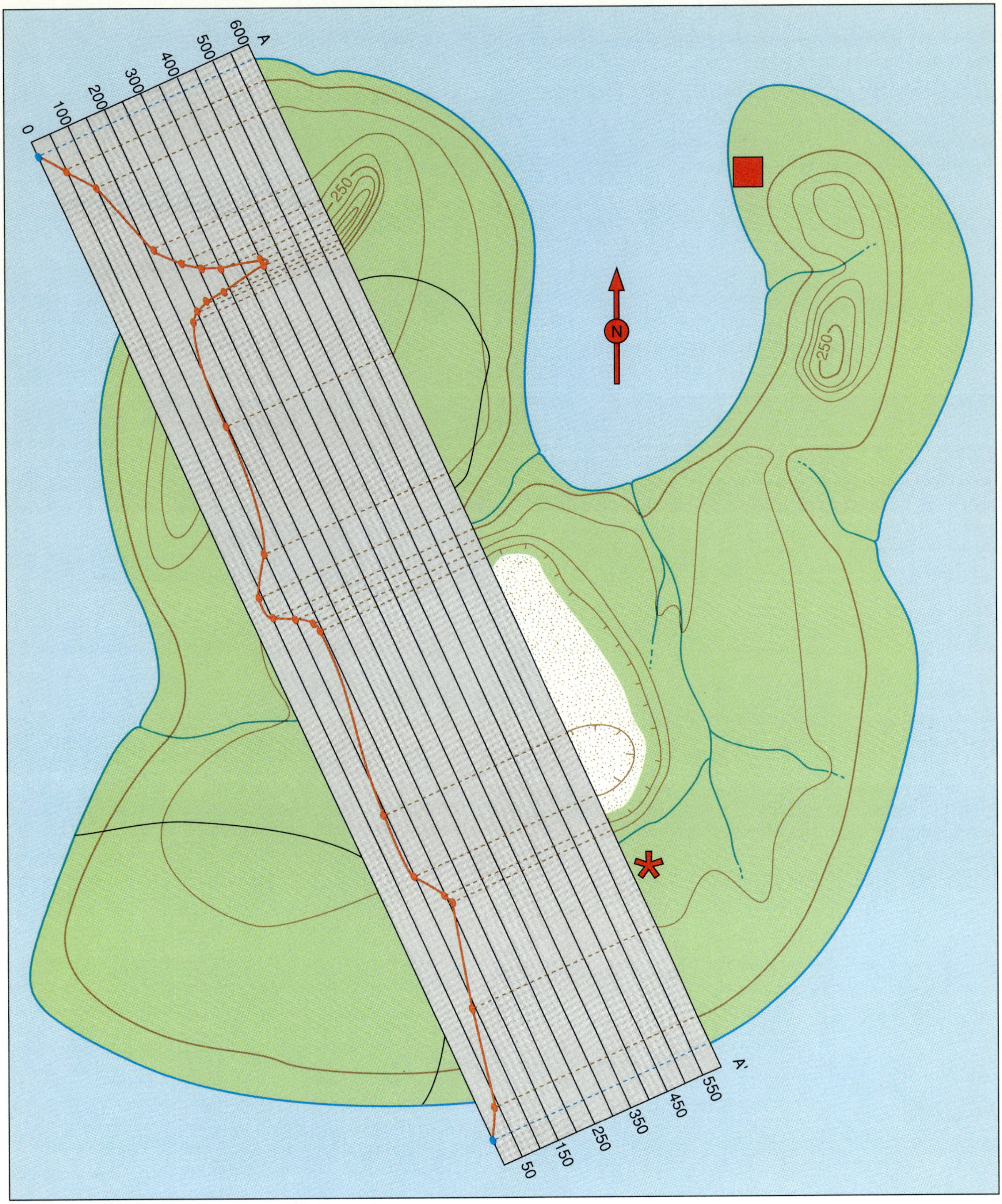

FIGURE 6.11 Constructing a Topographic Profile. (See text for instructions).

and wouldn't look like the formidable barrier that it is. A slight vertical exaggeration reveals subtle elevation changes that would be obscured without exaggeration. To determine the amount of vertical exaggeration in a profile:

1) Determine the proportional scale used for the profile.
2) Divide the proportional scale for the map by the proportional scale for the profile.

For example, if the map used a scale of 1:62,500, and your self-constructed grid used one inch to represent 1,000 feet (1:12,000), the vertical exaggeration would be:

$$\text{vertical exaggeration} = \frac{62{,}500}{12{,}000}$$
$$= 5.21, \text{ or approximately } 5\times.$$

What is the vertical exaggeration in the profile shown in figure 6.11?

Construct a profile that runs from the research center to the southernmost tip of the island. What vertical exaggeration did you use?

EXERCISE 6.7 MAKING A TOPOGRAPHIC MAP

You should now be able to read the topographic maps used in the following chapters to show how Earth's surface processes can be interpreted. There are some times when it is desirable to **make** a topographic map or other kind of map using contour lines (see, for example, Exercises 9.2 and 9.3 in chapter 9).

Figure 6.12 shows an area in which several points have been located whose exact elevations above sea level are known. In addition, the courses of two streams have been plotted. *Draw a contour map for this area that follows the rules you established earlier. Then draw a profile from the northeast corner of the area to the southwest corner.*

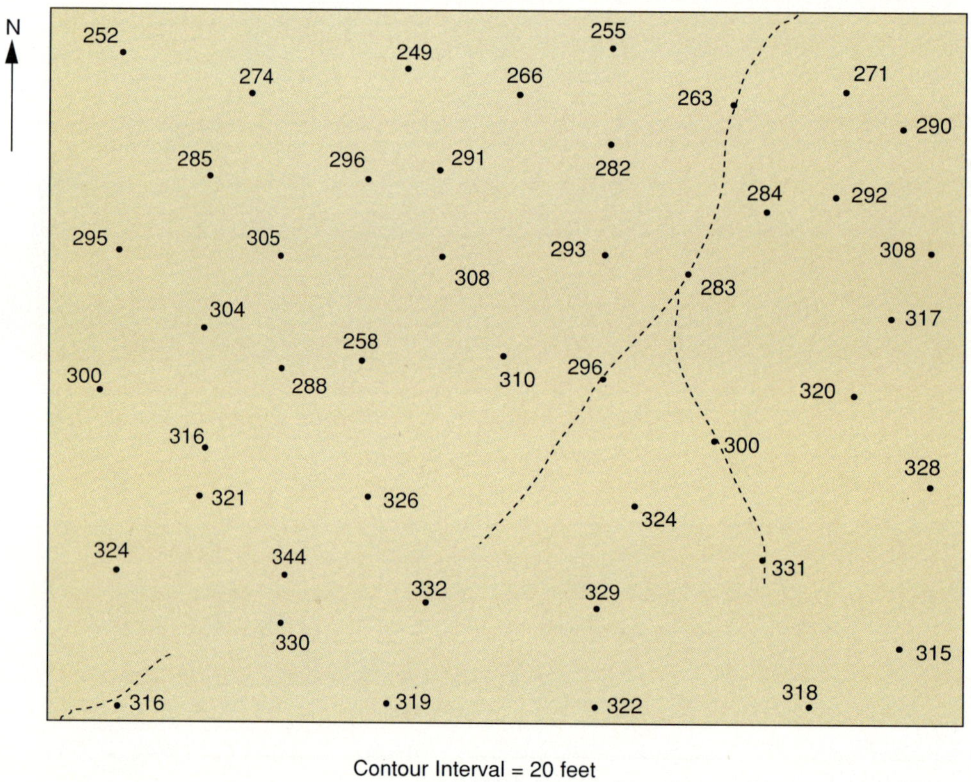

FIGURE 6.12 Surveyed points for a contour map.

EXERCISE 6.8 WHAT TO DO WHEN MEETING A TOPOGRAPHIC MAP

Topographic maps come in many different sizes and scales, cover different areas, and use many different contour intervals. Before even looking at the topography, look at the borders of the map. An enormous amount of helpful information is printed there, as shown in figure 6.13. Use figure 6.13 and the topographic map provided by your instructor to answer the following questions.

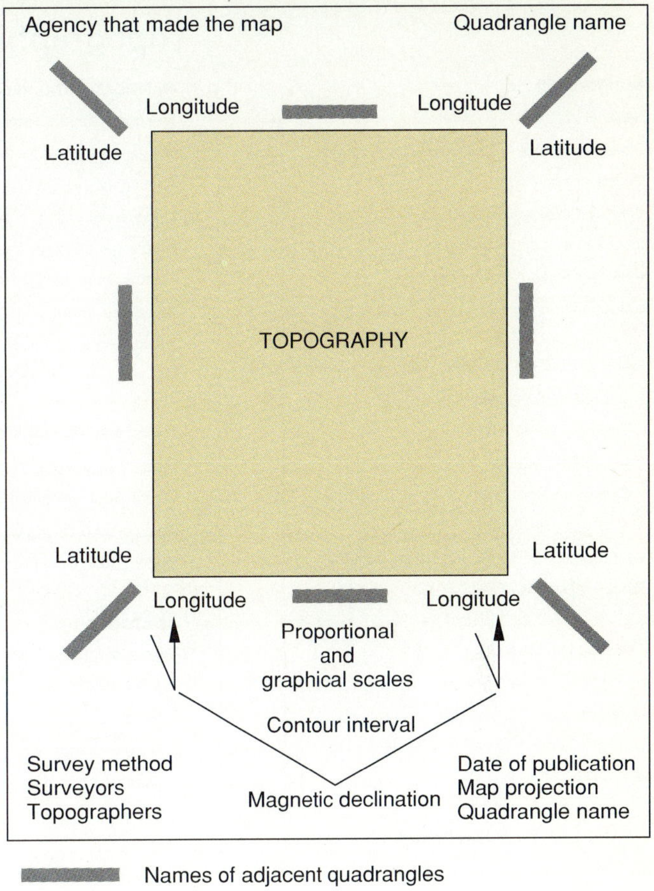

FIGURE 6.13 Vital statistics of a topographic map.

a) What is the scale of the map?

b) How much latitude (in degrees, minutes, seconds) does your map cover? How much longitude?

Topographic maps are called quadrangles, and are classified by how much latitude or longitude they cover. There are 7½-minute, 15-minute, 30-minute, 1-degree, and 2-degree quadrangles. *Which do you have?*

c) What is the magnetic declination in this area?

d) What are the names of the quadrangles that surround this one?

e) When was the map made? (This is vital, since an old map may not show human-induced changes.)

f) Was the map made by actual field surveying, or was it prepared from aerial photographs?

g) What projection was used to construct this map?

h) What is the contour interval?

The symbols commonly found on topographic maps are shown in figure 6.14.

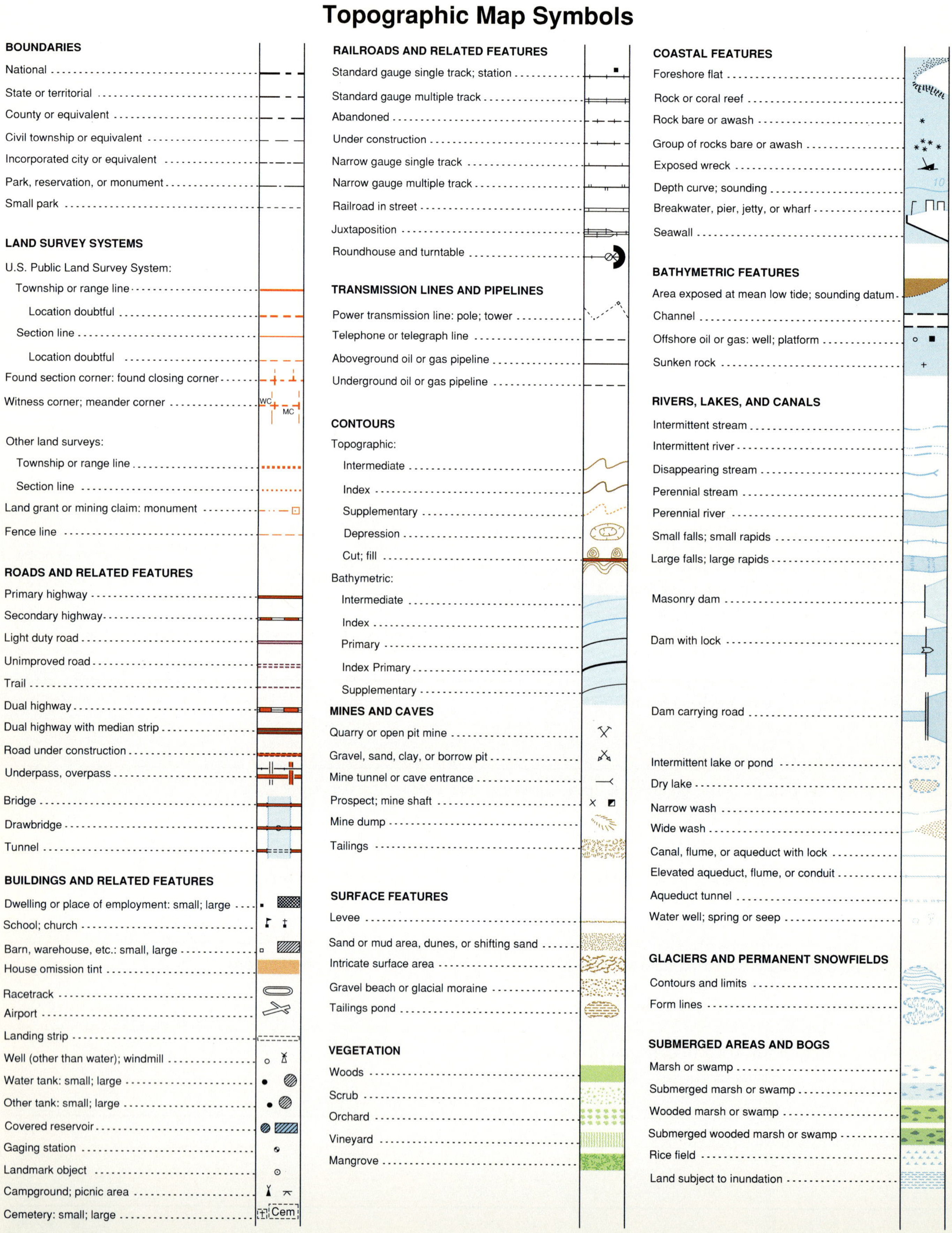

FIGURE 6.14 Topographic map symbols. (Source: U.S. Geological Survey.)

CHAPTER 7

STREAMS AND STREAM-CARVED LANDSCAPES

PURPOSE

To learn the factors that control the way streams operate; to become familiar with landforms that result from stream erosion and deposition.

MATERIALS NEEDED

1) A ruler, preferably one marked in tenths of an inch.
2) Thin string.
3) A magnifying glass or hand lens to help in reading closely spaced contour lines.

INTRODUCTION

Streams are the most effective agent of erosion and landscape formation on Earth. Even in deserts, where most people think wind is the dominant erosional agent, streams commonly erode more in a single storm than winds do in an entire year. Glaciers are also formidable erosional agents, but a glaciated landscape is very different, at least to the eyes of a geologist, than a stream-carved region. In the first part of this chapter, we will examine the ways in which streams operate, how they transport and deposit material. We will then look at how a landscape evolves as streams continue to cut their way through it, and how to distinguish stream-carved landscapes from other types.

Principles of Stream Erosion and Deposition

Regardless of size, all streams operate according to a simple set of principles, from the broad Mississippi River in Louisiana to the tiniest creek in Colorado, from the deeply entrenched Colorado River at the base of the Grand Canyon to the slowest moving run in Virginia or brook in Vermont. The scale is different for each, but the rules are the same:

Water in streams flows downhill because of the force of gravity.

The motion of the water gives a stream energy (kinetic energy) to do geologic work. The more energy a stream has, the more work it can do. The amount of energy depends on the amount of water and its velocity. The faster a stream flows and the more water it contains, the more energy it has. Thus, a large, slow-moving stream has less energy than a stream of equal size that is flowing faster.

Kinetic energy permits streams to transport sediment such as sand, silt, mud, and small boulders. These slide or roll along the bottom, bounce up into the stream, or are carried in suspension. When these particles bump against the solid bedrock walls of a stream channel, they chip away small pieces of rock. Millions and millions of such abrasions enable streams to erode their valleys through solid bedrock.

When a stream loses kinetic energy—by slowing down or evaporating—it can no longer carry all of its sediment load, and it begins to deposit material.

Exercise 7.1 Types of Streams

The rules may be the same, but their application, and thus the appearance of streams, varies widely. For example, compare figure 7.1a, the Yellowstone River in Wyoming, and figure 7.1b, the Cascapedia River in Quebec. These streams are quite different in the way they are going about their geologic work.

a)

b)

Figure 7.1 Two contrasting streams.

a) In your own words, what are the differences between these two streams? (Don't worry yet about the geologic terms; they will come soon enough.)

A stream **channel** is the area within which the water is actually flowing. A stream **valley** is the region within which the stream has eroded the land. In some cases, the channel fills the bottom of the valley completely, but in others, the valley is many times wider than the channel of the stream that carved it.

b) Which stream has the wider **valley?**

c) For which stream are the valley walls more distinctly developed?

d) Describe the relation between valley width and channel width for both streams.

e) Which stream seems to be flowing in a straighter course? Which has the more sinuous (winding) course?

f) Which stream seems to be flowing faster?

g) Which stream appears to be flowing more steeply downhill? (The steepness of a stream channel is called its **gradient,** and is a major factor in the nature of stream erosion.)

h) Which stream looks as if it has more energy?

i) Look carefully at the valley of the Cascapedia River. What evidence is there that at one time the river had more energy than it does now? Where could this energy have come from?

These questions delve into the controls on all stream behavior. All the variables are interrelated; a change in one causes changes in the others. For example, a change in gradient may completely change the nature of stream erosion, the width of its valley, and the degree to which its valley walls are clearly defined. Velocity changes lead to different patterns of erosion and deposition. It is important to keep these variables in mind when you look at photographs or maps of streams.

Let us try to correlate some of the variables with the appearance of the streams, based only on your observations. We will see later whether these correlations can be successfully generalized to all streams.

Gradient vs. Stream Sinuosity Is the steeper or gentler gradient associated with a straight stream course? With a sinuous course?

Channel/Valley Width vs. Gradient Which gradient is associated with the stream whose channel barely fills the base of its valley? Which is found in the channel that is much narrower than the valley?

Valley Wall Development vs. Channel Width The more sharply defined valley walls are associated with which kind of channel-to-valley width relation?

Stream Erosion: Downward or Sideways?

A stream is limited in how far downward it can erode its channel, because it can never carve lower than the elevation of its mouth. If it did, water would have to flow uphill to get to the end of the stream. Figure 7.2 shows how a stream's gradient changes as it erodes downward with time. The gradient is steepest early in the process, and becomes gentler as the headwaters of the stream are eroded closer to the level of the mouth. This lowest elevation to which a stream may erode is called its **base level.** The base level for streams that flow into the ocean is sea level. For streams that flow into other streams, base level is the elevation of the second stream at their junction.

When a stream has a high gradient, its energy is mostly used in carving **downward,** lowering its channel as rapidly as possible toward the base level. When the headwaters are lowered close to the base level, the stream still has kinetic energy to transport and erode, but it can no longer erode downward. At this point it begins to erode **laterally** (sideways) by meandering (Figure 7.2b), and eventually, it meanders even more widely (Figure 7.2).

The change from downward to sideways erosion causes a change from a V-shaped valley completely filled by the stream channel, to a broad-bottomed valley in which the channel occupies only a small part. The transition from mostly downward to mostly sideways erosion is marked in many streams by meanders that just fit the width of the valley; increased use of energy for lateral erosion widens the valley, and the meanders migrate across them.

It was once thought that all streams passed through three stages in their development: an early stage of downward erosion, followed by a combination of downward and lateral erosion, and finally lateral erosion. We now know that time is not a controlling factor in stream behavior. The gradient is often steeper in the headwater area of a stream than near its mouth. As a result, a single stream may be eroding downward in some parts of its course and sideways in others. In addition, a stream that flows across a gentle slope from the moment it first forms will have a gentle gradient from the beginning.

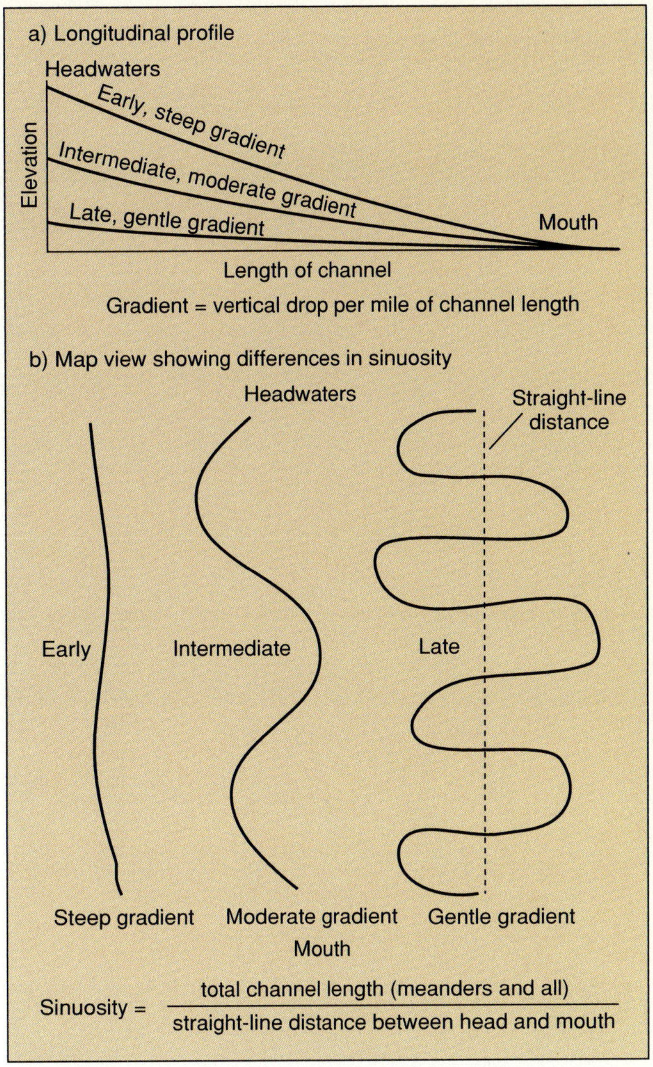

FIGURE 7.2 Changes in stream gradient and sinuosity over time.

EXERCISE 7.2 USING TOPOGRAPHIC MAP SKILLS TO ANALYZE STREAM BEHAVIOR

Photographs are useful for studying landforms, but they do not permit the detailed measurements and description that topographic maps allow. Carefully examine the Yellowstone (Figure 7.3), Mississippi (Figure 7.4), and San Jacinto (Figure 7.5) rivers on the Canyon, Lancaster, and Burnett Bay quadrangles, respectively.

First, establish the similarities and differences between the types of maps present.

a) What are the scales of each map?

b) What are the contour intervals used on each map?

FIGURE 7.3 Part of the Canyon, Wyoming Quadrangle. Scale: 1:125,000. Contour interval: 100′.

FIGURE 7.4 Part of the Lancaster, Wisconsin-Iowa-Illinois Quadrangle. Scale: 1:125,000. Contour interval: 20′.

FIGURE 7.5 Part of the Burnett Bay, Texas Quadrangle. Scale: 1:31,680. Contour interval: 1′.

Measure or determine the following features of the three rivers, and record them in table 7.1: gradient, valley width vs. channel width, sinuosity. Sketch a profile across each of the three to compare the valley shapes. To measure the length of a meandering stream (for both gradient and sinuosity), use a piece of string to trace the meanders. Then straighten the string and measure its length using the map scale. To determine the gradient of a river, first find out which way it is flowing. Determine the highest and lowest elevations along the channel by reading the contour lines—preferably those that cross the river. Then divide the maximum elevation change of the channel by the total channel length to compute the gradient in feet/mile.

Finally, based on what was discussed above about stream erosion, decide if each river is carving its valley mostly downward, sideways, or both.

TABLE 7.1 A COMPARISON OF THREE RIVERS

	RIVER		
FEATURE	**Yellowstone**	**Mississippi**	**San Jacinto**
Gradient			
Valley Width vs Channel Width			
Sinuosity			
Valley Shape (V-Shaped or Flat-Bottomed)			
Nature of Erosion (Downward vs Lateral)			

FEATURES FOUND WITHIN STREAM VALLEYS

The V-shaped valley of a stream that is eroding downward is due to two separate but related processes. The stream channel itself is carved downward by abrasion; by itself, this would produce a narrow valley with vertical walls. The widened top of the "V" results from gravity acting on overly steep slopes, causing slumping, landslides, and soil creep, which eventually result in the gentler slopes of the "V" (Figure 7.6a). Note that the stream itself does not carve the upper part of the V.

Features within the valley of a laterally eroding stream show how sideways erosion occurs, and how deposition takes place as well (Figure 7.6b). Meanders migrate with time, eroding at the outer, convex part of the meander loop and depositing material at the inner, concave part. This enables the channel to shift laterally, eroding more and more of the surrounding countryside without carving downward very far.

Widely meandering streams straighten their courses locally by cutting through meander loops, leaving former meanders isolated from the new channel. These are called **oxbows** and, when filled with groundwater, are known as **oxbow lakes.**

EXERCISE 7.3 EROSION AND DEPOSITION IN A RIVER VALLEY

Look again at the San Jacinto River in the Burnett Bay Quadrangle (Figure 7.5).

a) Locate places where former meander loops of the San Jacinto River have been cut off, leaving oxbows or oxbow lakes.

b) Describe the evolution of the valley walls on the east side of the river, particularly in the area just west of the cemetery on the west side of Crosby Road (north of Elena).

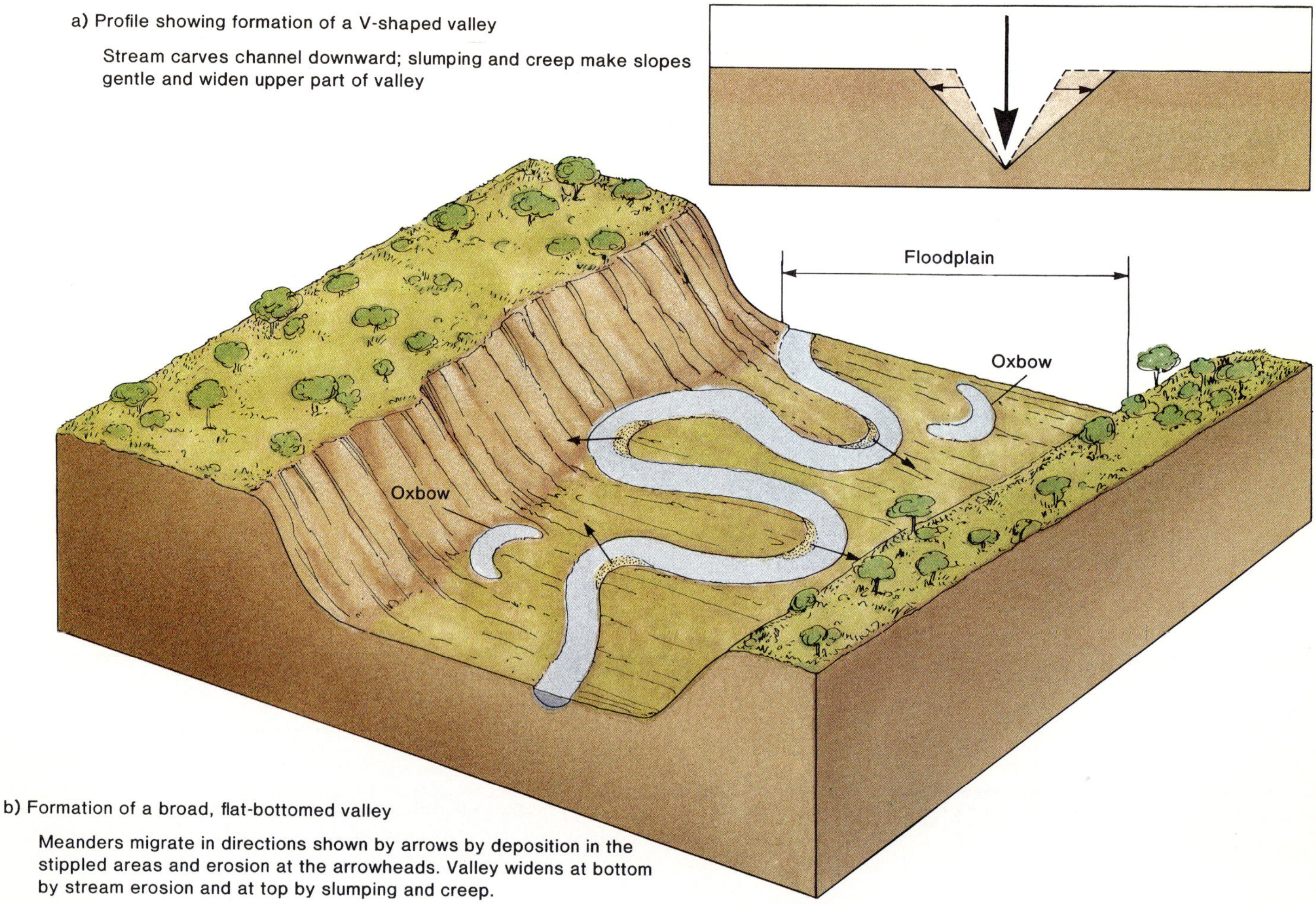

FIGURE 7.6 Development of stream valleys.

When a downward-eroding stream floods, its waters simply rise higher in the V-shaped valley that it occupies. When the San Jacinto River receives more water than its channel can hold (at times of heavy spring rains, for example), the water first fills the channel and then spreads out across the broad valley. The floodwaters deposit sediment across the width of the valley, producing what is called a **flood plain.** You might expect a flood plain to be perfectly flat, but the San Jacinto River shows that this is not always the case.

---

c) Draw a north-south profile across the San Jacinto River between the words Jacinto and River on figure 7.5.

Note the ridges that parallel the river. These form when a stream first overflows its banks during a flood, depositing large amounts of the coarsest sediment it is carrying. Such features are called **natural levees.**

---

d) Ideally, natural levees should be found at the margins of each meander loop, including former loops—the oxbows. Is this the case for the San Jacinto River?

Evolution of a Landscape Dissected by Streams

Just as a stream passes through stages of erosional behavior with time, an area that is being dissected by stream erosion also changes as more and more erosion occurs. The evolution of such a landscape involves changes in elevation, the number and gradient of its streams, and the amount of flat land compared with the amount of land on hillslopes.

Consider a large block of land that is uplifted high above its surroundings to form a plateau. What do you think will happen to it as streams become established, carve their way into it, and eventually remove most of it by erosion? To answer this question, just remember what happens to a stream as it carves its way down toward base level, and consider what would happen to an area in which many streams are developing over time. The following questions will help.

Exercise 7.4 Evolution of a Stream-Dissected Landscape

a) When will the area have its highest elevation (just after uplift or at the end of the erosional process)?

---

b) When would you expect streams to have their steepest gradients (when the erosion begins or when it is nearly completed)? Explain.

---

c) At what stage would you expect streams to be meandering widely across the block (at the beginning or at the end of the erosional cycle)? Explain.

---

d) At what stage would you expect there to be the greatest amount of hillslope present in the area (beginning, middle, end)? Explain.

Examine the topographic maps of the Soda Canyon, Colorado (Figure 7.7), Fayetteville, West Virginia (Figure 7.8), and Burnett Bay, Texas (an old friend by now—figure 7.5) quadrangles. Each represents a different stage of the stream erosional cycle of an area. Answer the following questions about the three areas by filling in table 7.2:

FIGURE 7.7 Part of the Soda Canyon, Colorado Quadrangle. Scale: 1:62,500. Contour interval: 50′.

FIGURE 7.8 Part of the Fayetteville, West Virginia Quadrangle. Scale: 1:62,500. Contour interval: 50′.

TABLE 7.2 STAGES OF STREAM DISSECTION

	SODA CANYON	FAYETTEVILLE	BURNETT BAY
Land shape: Mostly flat			
Mostly hilly			
Stream divides: Broad, flat			
Narrow, sharp			
Relief: Highest elevation			
Lowest elevation			
Total relief			
Streams: Straight			
Meandering			
Gradients			
Downcutting			
Side-cutting			
Both			
Stage of dissection			

---

e) What are the highest elevations in each area?

In which area do the major streams have the steepest gradients?

What is the shape of the stream divides—the land surface that separates streams: broad and flat, or narrow and sharp?

In which area(s) are streams mostly straight? Meandering slightly? Meandering widely?

In which area(s) are streams eroding mostly downward? Sideways? Both?

In which area(s) is the land mostly flat? Mostly hilly?

The three stages of stream dissection are commonly referred to as youth (early stage), maturity (middle stage), and old age (final stage).

---

f) In which stage is each of the three areas? Cite evidence for your conclusions.

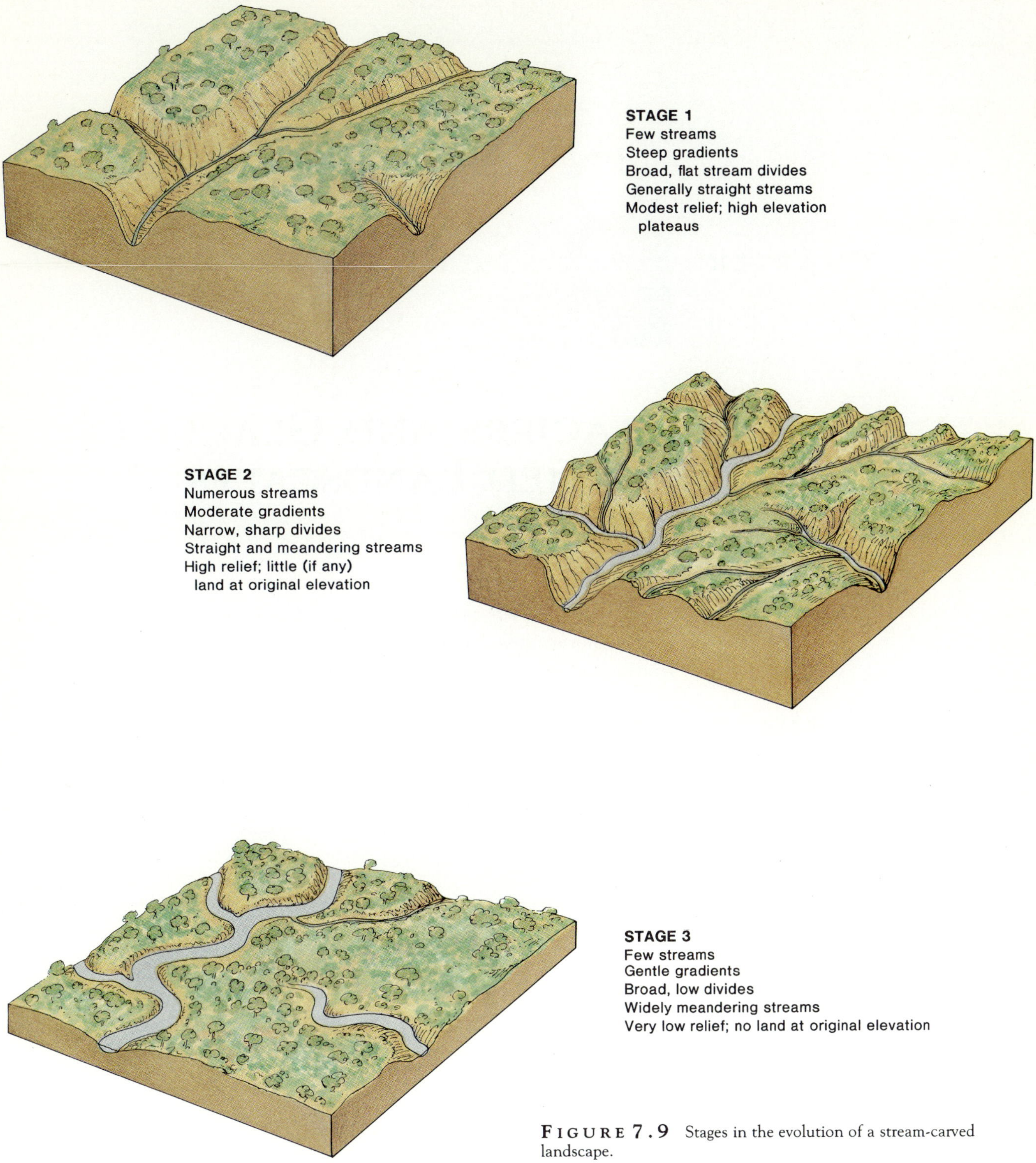

FIGURE 7.9 Stages in the evolution of a stream-carved landscape.

---

g) Figure 7.9 summarizes the observations made on the previous page. Label each of the three blocks with the name of the corresponding quadrangle.

CHAPTER 8

GLACIERS AND GLACIALLY-FORMED LANDSCAPES

PURPOSE

To learn how glaciers carry out erosion and deposition; to become familiar with landforms carved and deposited by glaciers; to be able to distinguish landscapes formed by glaciers from those produced by stream action.

MATERIALS NEEDED

1) A magnifying glass or hand lens to help read close-spaced contour lines.
2) Colored pencils.

INTRODUCTION

It is hard for most of us who live in temperate or tropical climates to realize that the amount of water frozen in Earth's glaciers is more than 21,000 times that found in all the streams on the planet. Today, huge continental ice caps cover most of one entire continent—Antarctica—and nearly all of the misnamed island of Greenland. Smaller glaciers occupy depressions and valleys in the highest mountains and are responsible for most of the familiar topography of the Alps, Himalayas, and Northern Rocky Mountains.

Despite this enormous volume of ice, glaciers contain only a small percentage of Earth's water—most, by far, is in the oceans. The balance in the water cycle between glaciers, oceans, streams, and groundwater changes during periods of increased glaciation (Ice Ages). This affects the base levels of streams (chapter 7) and the nature of shoreline processes as well (chapter 11).

In Earth's recent past, enormous ice sheets formed several times in the northern hemisphere and spread over much of North America and Europe. Erosional and depositional features from these "Ice Ages" form the landscape in most of Canada and the northern United States. It is obvious to a trained eye that these landforms could not be the result of stream activity. The differences result from the different ways in which glaciers and streams go about their geologic work.

EXERCISE 8.1 COMPARISON OF GLACIATED AND STREAM-DISSECTED AREAS

Your eye is already halfway educated, having become familiar with landscapes formed by streams. Look at figure 8.1, a photograph of an area in Glacier National Park, Montana. Compare the valley in the background with the valleys of the Yellowstone and Cascapedia rivers (Figure 7.1a, b), and with the valley profiles shown in figure 7.6.

a) How is the shape of this valley different from those that we know are carved by streams?

b) This shape is characteristic of glacially carved valleys. Can you think of a reason for the difference? (We will help in a minute.)

FIGURE 8.1 A valley in Glacier National Park, Montana. Compare this glacially carved valley with stream-cut valleys in figure 7.1.

PRINCIPLES OF GLACIAL EROSION AND DEPOSITION

Although the ice in a glacier is solid, gravity acting on a thick mass of ice causes it to behave like a very sluggish fluid, particularly near its base. Glaciers thus flow (like a stream), but their solidity gives them properties alien to streams. Some of the similarities and differences between streams and glaciers include:

The flow of a glacier provides kinetic energy, as does the flow of water in a stream. Glaciers flow more slowly than streams but have much greater mass and thus enormous kinetic energy.

The solidity of a glacier gives it the structural strength to transport particles far larger than those carried by streams—blocks weighing hundreds or even thousands of tons.

Like streams, glaciers erode by abrasion. Particles carried in the ice scrape against bedrock and scour it away. Unlike streams, the particles don't bounce as they move; they are frozen into the ice and behave like the grains on a sheet of sandpaper.

Like streams, glaciers deposit sediment when they lose kinetic energy. This happens not when they slow down, however, but when they lose mass by melting.

Glaciers form in snowfields where more snow falls each year than melts. Gravity acting on the snowfield creates pressure that converts snow to ice and then forces the ice outward in all directions. This causes the glacier to extend to areas well beyond the snowfield.

Glaciers do not have a base level. The depth to which a glacier carves a valley on the land depends on the amount of ice, its rate of flow, and the amount of abrasive material it carries. The bigger and faster a glacier is, the more deeply it can erode.

U-Shaped vs V-Shaped Valleys Figure 8.2 answers the question raised earlier about why glaciers carve U-shaped valleys rather than the V-shaped or broad, flat valleys eroded by streams. This single topographic characteristic is enough to determine which of the two agents of erosion was responsible for landscape development in a given area. Figure 8.2a shows how a downward-cutting stream creates a valley; active erosion is accomplished in the stream channel, **only at the very bottom of the valley.** In contrast, a glacier may fill the entire valley (Figure 8.2b); sediment frozen into the ice can carve into bedrock **everywhere the ice is in contact with rock.** The glacier carves downward **and** laterally at the same time, everywhere in the valley it fills, not just at the bottom.

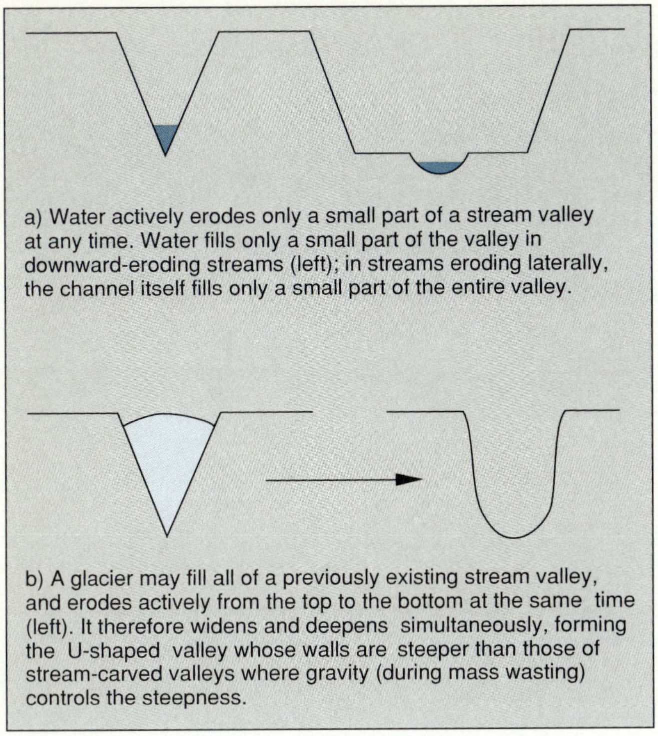

FIGURE 8.2 Comparison of glacial and fluvial valley formation.

Types of Glaciers

Glaciers can be grouped into two main categories. **Alpine (valley) glaciers** form in relatively small snowfields in depressions on mountainsides; they flow downhill, filling and modifying previously existing stream valleys. **Continental glaciers (ice sheets, ice caps)** develop in huge snowfields on continents and spread outward radially in all directions. Alpine glaciers may be hundreds of feet thick; continental glaciers are often more than a mile thick. The different scales result in somewhat different landforms. Alpine glaciers modify previously existing topography but leave evidence of the former landscape. Continental glaciers overpower everything in their way and may fill in old rivers, causing them to disappear.

Exercise 8.2 Alpine Glaciation

Alpine glaciers produce a dramatic, unmistakable landscape filled with distinctive erosional features. Examine figure 8.3, a portion of the Chief Mountain Quadrangle in Montana. Note that a few small alpine glaciers are left in the higher elevations; there is plenty of evidence that such glaciers covered nearly all of the area at one time.

a) Aside from the glaciers themselves, there are several prominent examples of another feature that was absent from the stream-carved landscapes you studied in chapter 7. What are these features? Describe their location and relation to one another.

b) We saw in chapter 7 that streams have smooth, concave-up longitudinal profiles (from source to mouth). Briefly sketch and describe the longitudinal profile of the unnamed stream that flows eastward through Twin Lakes to St. Mary River in the southern part of the map. How does it compare with the longitudinal profile expected of a stream-carved valley?

FIGURE 8.3 Part of the Chief Mountain, Montana Quadrangle. Scale: 1:125,000. Contour interval: 100′.

c) Construct or sketch a profile from Kennedy Lake (near the eastern edge of the map) northwestward across the Ptarmigan Wall to the unnamed hill with an elevation of 7601 feet.

What shape is the valley of the Belly River?

Note that the Belly River seems very small compared to the width of its valley. This is another characteristic of glacially eroded areas: glaciers carve large valleys, which are later used by streams that are too small to have eroded them.

How did the Ptarmigan Wall form? (HINT: What erosional agent has been active on either side of it?)

A knife-sharp ridge like the Ptarmigan Wall that separates two U-shaped valleys is called an **arête,** and is a common feature in alpine glaciation.

Locate other arêtes in the map area.

---

d) There are two kinds of lakes in the area. Some are nearly circular, while others are elongate. Round lakes (e.g., Ptarmigan and Kennedy lakes in the northeast corner of the map) occur in different topographic positions than elongate ones, like Glenns Lake, near the center of the map. *Describe the differences in topographic location.*

Alpine glaciers form in depressions high on mountain sides. They eventually widen and deepen these depressions, creating basins called **cirques.** When the ice melts, the basins sometimes fill with water to form lakes called **tarns.**

Which of the different types of lakes are tarns (round or elongate)?

Locate cirques that have not been filled to form tarns.

Knowing what you do about the typical longitudinal profile of a glaciated valley (see Exercise 8.2b), suggest an origin for the elongate lakes.

Another feature common to alpine glaciation is a steep-sided, sharp-pointed mountain called a **horn peak** (or simply a **horn**), after the most famous example, the Matterhorn in the Alps. Horns form when erosion of several cirques on the slopes of a more rounded hill eats into the hill, leaving the classic horn shape.

---

e) Locate and give the names of several horn peaks in the area.

Waterfalls are also abundant in areas affected by alpine glaciation. Waterfalls are rare in normal stream erosion for two reasons: (1) streams have smooth gradients, and (2) a main stream serves as base level for a tributary stream.

f) Knowing what you do about factors that control the depth to which glaciers can carve their valleys, explain the origin of waterfalls such as Appekunny Falls (eastern part of the map where Appekunny Creek enters Lake Sherburne).

Horns, arêtes, cirques, underfit streams (those that are much too small for their valleys), U-shaped valleys, tarns, and finger-shaped lakes are all indicative of alpine glaciation. Briefly look over the maps in chapter 7 to reinforce how different typical stream-carved landscapes are from those formed by alpine glaciation.

Continental Glaciation

Large river valleys filled by continental glaciers display the typical U shape of all glaciated valleys, but cirques and other erosional features associated with alpine glaciation do not develop. Continental glaciers do not just flow through previously existing valleys; they move as walls of ice that cover the ground, bulldozing loose sediment and abrading the entire land surface. Figure 8.4 shows how a continental glacier modifies the shape of bedrock hills. Glaciers produce distinctively asymmetric hills; the gentler slope faces the direction from which the glacier came; the steeper side faces the direction toward which it was flowing.

In some cases, water may be present at the base of a glacier, while in others, there may be no water at all, and the glacier is instead frozen to the ground below. In the latter case, the flow of the ice rips up pieces of the underlying rock material, a process called **plucking.** The plucked material is then added to the sediment load of the glacier.

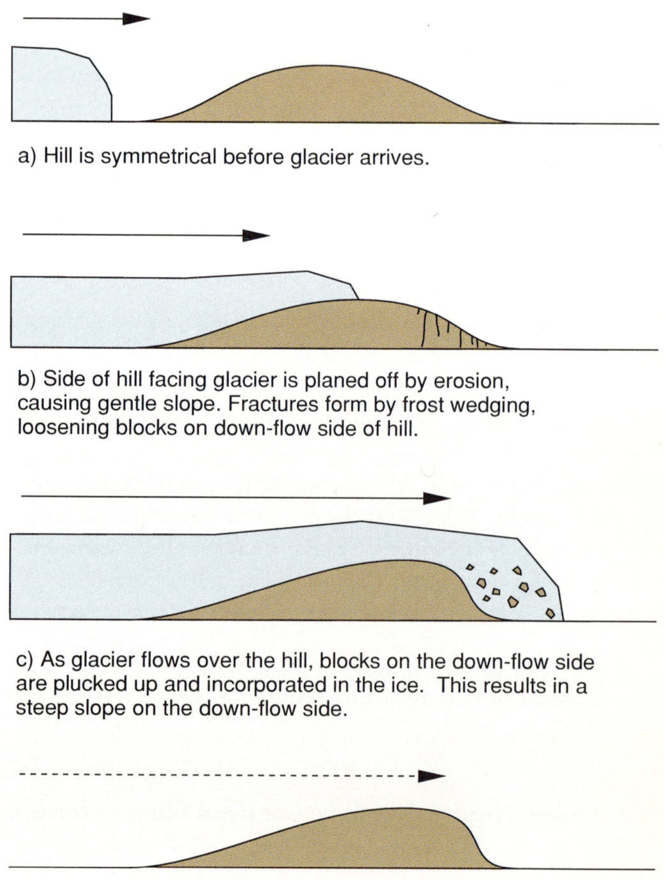

a) Hill is symmetrical before glacier arrives.

b) Side of hill facing glacier is planed off by erosion, causing gentle slope. Fractures form by frost wedging, loosening blocks on down-flow side of hill.

c) As glacier flows over the hill, blocks on the down-flow side are plucked up and incorporated in the ice. This results in a steep slope on the down-flow side.

d) After ice melts, its flow direction is indicated by the asymmetric hill.

FIGURE 8.4 Modification of hills by a continental glacier.

Exercise 8.3 Erosion by Continental Ice Sheets

Examine the portion of the Peach Lake, New York, Quadrangle shown in figure 8.5. The hills are supported by high-grade metamorphic rocks, mostly gneisses, that are more resistant to erosion than adjacent marbles and schists.

FIGURE 8.5 Part of the Peach Lake, New York Quadrangle. Scale: 1:24,000. Contour interval: 10′.

a) Sketch or construct a north-south profile from Saint John's Church to Lake Rippowam.

b) Using your colored pencils, color-code the steep sides with one color and the gentle sides with another. Is there a consistent orientation of the steep and gentle slopes of the asymmetric hills? Describe this orientation.

c) Draw arrows to indicate the direction in which the glacier flowed across the map area.

Landforms Produced by Glacial Deposition

All glaciers deposit their sediment load when they melt. The mud, silt, sand, pebbles, boulders, and blocks up to half a kilometer long are dropped where the ice melts, leaving a poorly sorted jumble of material called **till.** Water from the melting ice forms streams that pick up the finer grained particles and redistribute them. These materials, called **outwash,** are much more uniform in grain size than till, because they are sorted by the streams.

Landforms composed of till are called **moraines.** When a glacier melts back steadily (recedes), it deposits a sheet of till that blankets the entire area formerly covered by the ice: a **ground moraine.** When the front of the glacier stays in one place where till can be delivered over a long period of time, a ridge of till forms that outlines the edge of the glacier. This is called a **terminal** or **end moraine.** Ground and terminal moraines are highly irregular features, pitted with many depressions that once held blocks of ice that did not immediately melt with the rest of the glacier. These small basins are called **kettle holes,** and sometimes fill with water to form kettle lakes. The only difference between moraines produced by alpine and continental glaciation is their size. Figure 8.6 shows terminal (in the foreground) and ground moraines deposited as the alpine glacier in the background retreated. Figure 8.7 outlines the extent of terminal moraines deposited by continental glaciers during the last major ice advance in Pleistocene times. Note that all of Long Island, Martha's Vineyard, and Cape Cod belong to two major moraines.

Outwash deposits follow the rules of stream deposition, and generally form sheets called **outwash plains.** Since the glacier itself did not cover the area of its outwash plain, there were no isolated blocks of ice to form kettle holes; as a result, outwash plains are much smoother than ground or terminal moraines. Not all outwash forms outwash plains, however. Tunnels filled with flowing meltwater often form within large glaciers. Sediment is sorted by the water, and when the ice melts, the shape of the tunnel is preserved in the midst of the ground moraine by an elongate (often meandering) ridge called an **esker.**

FIGURE 8.6 Deposition of till to form moraines. Three different types of moraines are visible here, along with the glacier that deposited them. The ridge in the foreground is a terminal moraine that marks the furthest extent of the glacier. The blanket of debris in the middle ground is a ground moraine. Low ridges at the sides of the valley are lateral moraines. (Juneau Ice Field, Alaska. Photography by Mitchell Algus.)

FIGURE 8.7 Southern extent of glaciation in eastern North America as shown by terminal moraines.

N
9.0
10.0
11.0
12.0
13.0
14.0
18.0
Roulier
Harricana
Saint-Narcisse
QUEBEC
Ottawa
River
OTTAWA
MONTREAL
St. Lawrence
Lake Champlain
MAINE
Pineo Ridge
GULF OF MAINE
Cape Cod
Sandwich
Nantucket
VERMONT
N.H.
MASS.
BOSTON
R.I.
CONN.
Charlestown
Harbor Hill
Ronkonkoma
NEW YORK
Hudson River
Dummer
Oak Ridges
LAKE ONTARIO
TORONTO
Albion
Hamburg
HAMILTON
Galt
Valley Heads
Escarpment
Defiance
Kent
PENNSYLVANIA
NEW JERSEY
MARYLAND
DELAWARE
WEST VIRGINIA
VIRGINIA
80°
76°
72°

EXERCISE 8.4 RECOGNIZING LANDSCAPES FORMED BY GLACIAL DEPOSITION

Examine the Whitewater, Wisconsin, map area shown in figure 8.8. This is an area that was covered by continental glaciers during the last major advance of ice across North America.

a) Describe the topography in your own words (we will look for specific geologic features shortly).

b) Compare this topography with those carved by streams. How is it different? What features are present here that are not found in areas affected only by streams?

c) A linear topographic high runs across the map from northeast to southwest. Suggest an origin for this feature.

d) Numerous hachured contour lines are found throughout the map, but are particularly abundant on this ridge. What features do they represent?

e) How are the lakes shown here different in origin from those formed during alpine glaciation that you investigated in Exercise 8.2?

The southeastern corner of the map is located at approximately 42°45′ N, 88°35′ W. Locate the area on the glacial map shown in figure 8.7. This lets you see how this specific area was related to major glacial features during the last Ice Age, and helps you understand the nature of other features on the map.

f) The topography northwest of the linear ridge is quite different from that of the ridge itself. What kind of feature is represented?

g) Streams like the Scuppernong River are beginning to erode through the glacial deposits, but have not yet fully developed the drainage patterns characteristic of stream-dissected areas. This can be explained by the interaction between glacial and fluvial (stream) processes. Explain! (HINT: the origin of the swampland throughout the low-lying area is an important clue.)

h) Use your colored pencils to outline 1) the farthest extent of glacial ice, as evidenced by your analysis of topography and figure 8.7; 2) areas underlain by different types of glacial deposits (color them differently and label them).

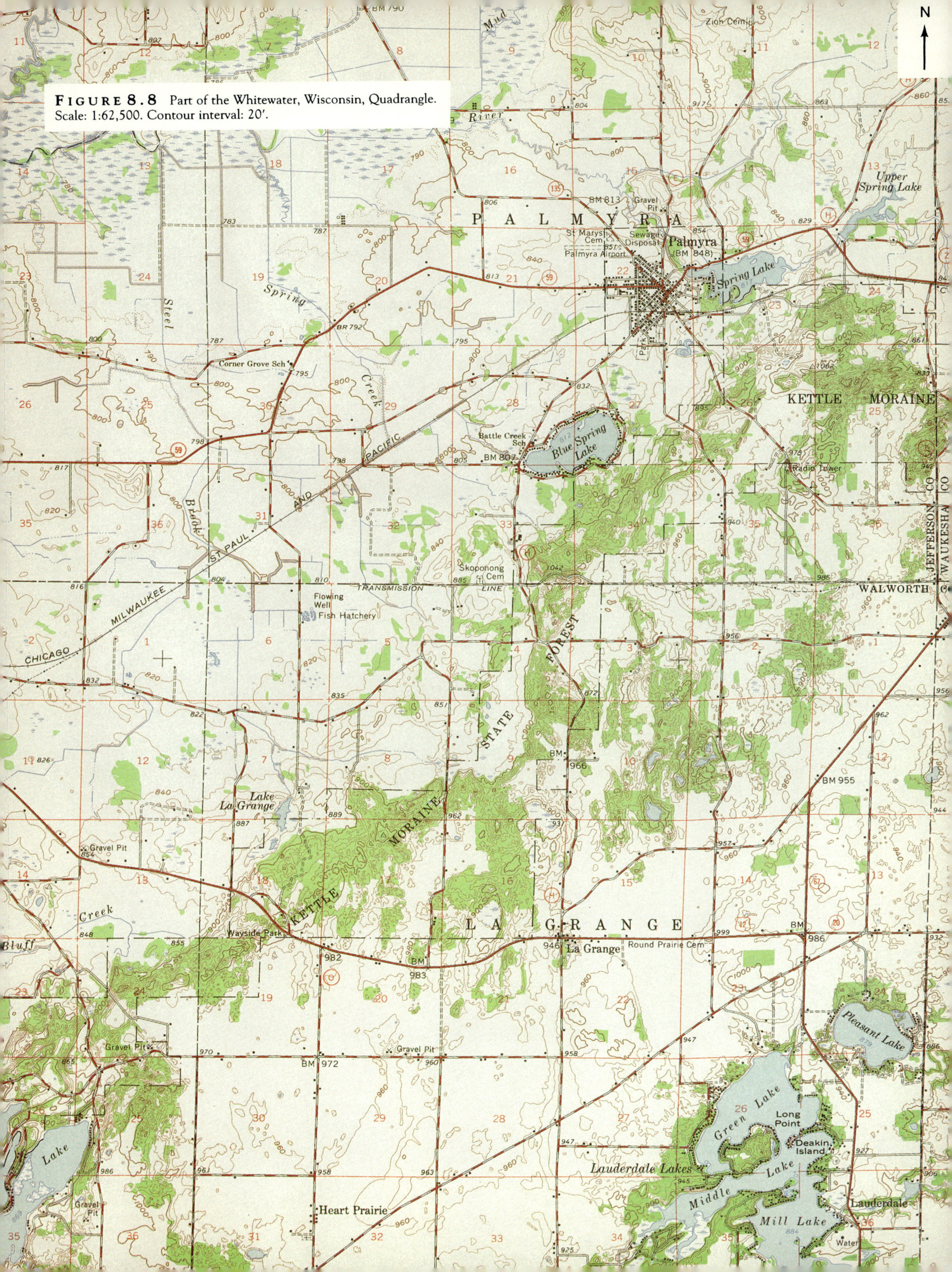

FIGURE 8.8 Part of the Whitewater, Wisconsin, Quadrangle. Scale: 1:62,500. Contour interval: 20′.

FIGURE 8.9 Part of the Passadumkeag, Maine, Quadrangle. Scale: 1:62,500. Contour interval: 20′.

Now look at the portion of the Passadumkeag, Maine, Quadrangle shown in figure 8.9

---

i) Locate Maine on the glacial map (Figure 8.7). What would you expect the major glacial depositional feature in the area to be? Explain.

---

j) Describe the landforms present. Is the general topography what you would have expected? Explain.

---

k) Two narrow, sinuous ridges are present. Compare these to the ridge in the Whitewater Quadrangle (Figure 8.8). How are they different?

---

l) What are these features, and how did they form?

---

m) Based on their heights and widths, give as much detail as possible about the glacial process by which they formed.

Look at the portion of the Weedsport, New York, Quadrangle shown in figure 8.10. We will use the features here to do a little bit of geological detective work.

---

n) Describe the topography in your own words.

---

o) With a colored pencil, sketch the long axis of one of the elongate hills. Now do the same for the others. What is the relation between these lines?

---

p) Sketch or construct a profile along one of these axes. Are these hills symmetrical or asymmetrical? Describe.

---

q) Compare these hills in size and shape to those in figure 8.5. Are these the same kind of "sculptured rock"? Why? (Or why not?)

Now that you've had a try at that, here is a clue: these hills are not composed of bedrock, but rather are made of till, and are called **drumlins.** (A brief historical note: Bunker Hill, of Revolutionary War fame, is a drumlin. Breed's Hill, on which the Battle of "Bunker Hill" was actually fought, is also a drumlin.)

FIGURE 8.10 Part of the Weedsport, New York, Quadrangle. Scale: 1:62,500. Contour interval: 10′.

r) Based on the glacial map in figure 8.7, in what direction did the ice move as it passed through this area?

s) Based on the orientation of the long axes of the hills, in what precise direction did the ice move? Show this with a large arrow on the map, and with an arrow on the profile that you sketched.

t) Compare the asymmetry of drumlins and the direction of associated glacial advance with the same aspects of the sculptured rock in figure 8.4, analyzed in Exercise 8.3.

Drumlins are among the best pieces of evidence for the precise direction of glacial advance, but it is absolutely critical to know whether you are looking at a drumlin or at glacially sculptured bedrock. *Explain why.*

CHAPTER 9

GROUNDWATER AS A LANDSCAPE-FORMER AND A RESOURCE

PURPOSE

To become familiar with the landscapes formed by groundwater erosion, and to be able to distinguish them from fluvial and glacial landscapes; to use geological skills in solving environmental problems related to our use of groundwater as a resource.

MATERIALS NEEDED

1) Colored pencils.
2) Brains and a pencil (or a calculator) to do some simple arithmetic.
3) A magnifying glass or hand lens to help read close-spaced contour lines.
4) Tracing paper.

Introduction

Water that sinks into the ground after a rain or when snow and ice melt is called **groundwater.** It does not flow through open channels like stream water, but moves very slowly from one tiny pore space to another. It cannot carry sediment particles through these tiny openings, and thus cannot erode by abrasion the way streams and glaciers do. Instead, groundwater erodes by **dissolving** those materials that are particularly soluble: rock salt and gypsum, but most commonly limestone, dolostone, and marble.

Groundwater is a vital natural resource as well as an agent of landscape formation. Every well drilled to find water is looking for groundwater. Rivers and reservoirs supply about half of the water we use for drinking, bathing, and many industrial purposes; groundwater provides the other half. Knowledge of how groundwater moves can help us avoid environmental problems such as pollution of our drinking supply by industrial contaminants or household waste. The following exercises deal with both aspects of groundwater.

Landscapes Carved by Groundwater

You might be wondering how groundwater, since it is beneath the surface of the Earth, can create landscape features. Figure 9.1 shows how underground erosion leads to a unique type of landscape called **karst topography.** Water percolates slowly into the ground through interconnected pores, but can move more rapidly along near-vertical fractures in bedrock. Dissolution of the soluble rock around these fractures causes a bowl-shaped depression at the surface, called a **sinkhole.** These fractures may become major conduits for groundwater, funneling it downward, where less steeply inclined surfaces—other fractures or bedding planes—also control its flow. In some cases, an underground cavity is hollowed out by the solution process,

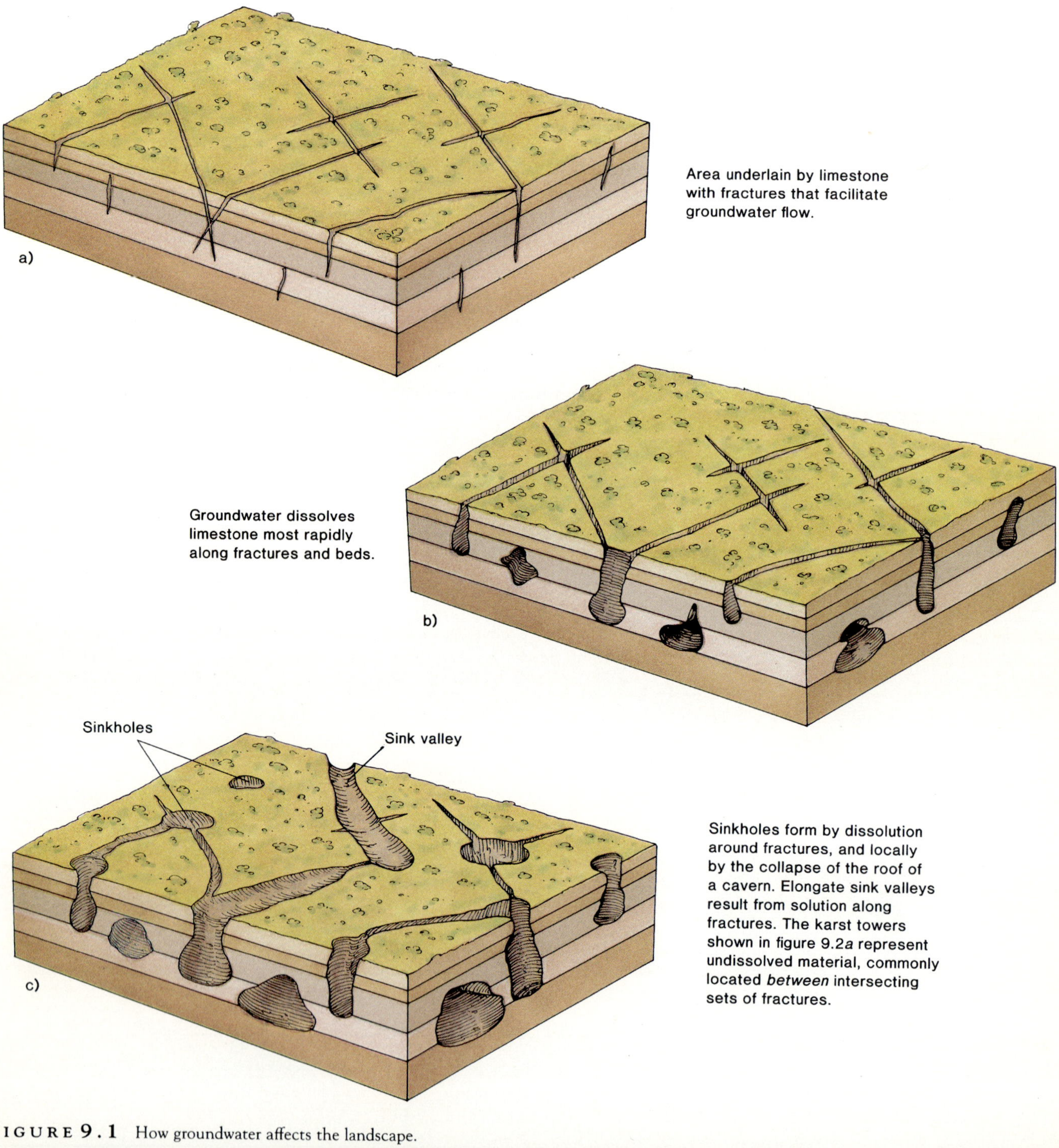

FIGURE 9.1 How groundwater affects the landscape.

a) Karst topography near Guilin, southeastern China

b) Karst topography in the midwestern United States

FIGURE 9.2 Karst topography.

forming spectacular caverns such as those at Mammoth Cave in Kentucky, Carlsbad (New Mexico), and Luray (Virginia) Caverns. When a cavity undermines its roof so that the overlying rock can no longer be supported, the roof collapses, also resulting in a sinkhole.

Karst topography is named after the area in Yugoslavia where it was first described, but sinkholes, sink valleys, caverns, and even spectacular karst towers (Figure 9.2a) can develop wherever rocks are highly soluble and there is an abundant supply of water. Extensive areas of karst topography are well developed in Yugoslavia, China, Puerto Rico, Jamaica, and parts of Kentucky, Indiana, and Florida.

EXERCISE 9.1 FEATURES OF KARST TOPOGRAPHY

Figure 9.2a is a photograph showing part of a large karst topographic region in southern China. The features seen here are typical of areas in which extensive groundwater solution has taken place over a long period of time.

a) Describe this landscape in your own words. Compare its hills and valleys with those of fluvial and glacial landscapes. What makes it seem so different? (HINT: It may help for you to quickly sketch what you think a simple contour map of one of these hills might look like, then what several would look like on a topographic map. Then compare with the maps in preceding chapters.)

(A brief cultural note: For many years, westerners thought that Chinese scroll painters were not able to accurately portray landscapes because the hills in their paintings looked so different from those of Europe and North America. The painters turned out to be quite accurate! The hills *are* very different.)

b) Now that you know something of how groundwater erodes, suggest an origin for the "strangeness" of these features.

Examine the portion of the Mammoth Cave, Kentucky, Quadrangle shown in figures 9.3 and 9.2b. Karst topography is in evidence here, although it is not as striking as that shown in figure 9.2a.

---

c) Before looking at any feature in detail, scan the entire map. Note that the topographic "grain" changes across the area shown. With a colored pencil, draw lines to separate places where the "grain" is different.

---

d) This part of the United States is quite humid, yet there do not seem to be many streams around. Suggest an explanation.

---

e) Describe the topography in the southern third of the map area. How did it form?

The hachured contour lines outline sinkholes, which formed by the mechanism described earlier and in figure 9.1.

---

f) Now look at the central third of the map area. Sinkholes are present here also, but are they the same as those in the south? Describe the similarities and differences, and suggest an explanation.

---

g) There are no sinkholes in the northernmost part of the map. Why not?

---

h) What special problems would you face if you wanted to build a skyscraper in this area?

Now look at the portion of the Interlachen, Florida, Quadrangle shown in figure 9.4. This is another karst area underlain by highly soluble limestone, and characterized by the absence of a stream network. The hachured contour lines here also indicate sinkholes, but these are different from those in the Mammoth Cave area because many have filled with water, forming lakes. A brief study of these sinkholes reveals an important aspect of groundwater.

---

i) Construct an accurate topographic profile along the line indicated on the map.

FIGURE 9.3 Part of the Mammoth Cave, Kentucky Quadrangle. Scale: 1:62,500. Contour interval: 20′.

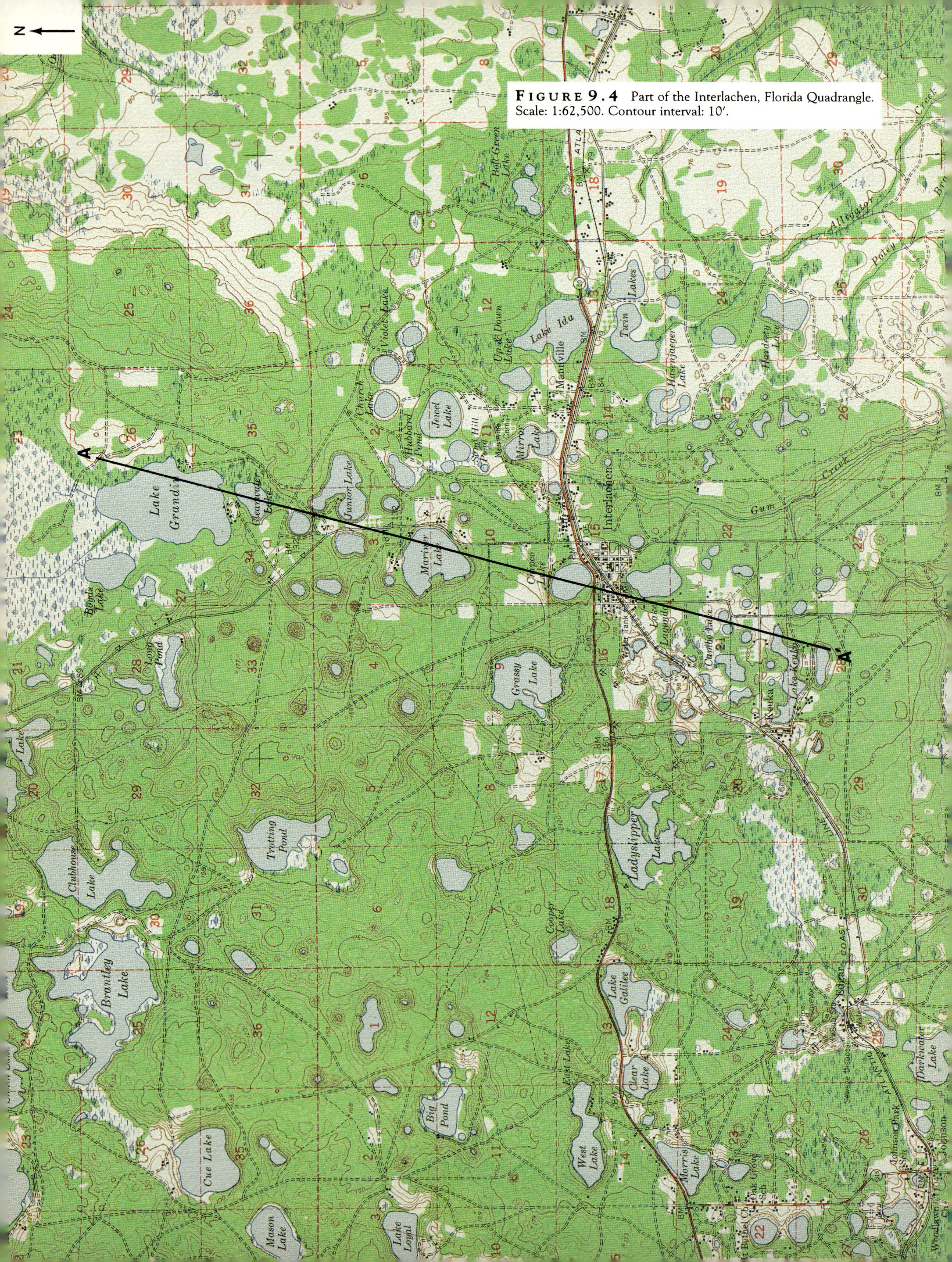

Figure 9.4 Part of the Interlachen, Florida Quadrangle. Scale: 1:62,500. Contour interval: 10′.

j) Some of the sinkholes contain lakes, and others are dry. Why? (HINT: After you have completed your profile, use a blue pencil to indicate the water level in each sinkhole that houses a lake. How does this level compare with the elevations in the dry sinkholes?)

The level of the lakes in the sinkholes is controlled by a feature called the **water table.** When water sinks into the ground, whether into unconsolidated sediment or bedrock, it moves downward until it reaches an impermeable barrier through which it cannot flow. Additional water percolating downward has nowhere to go, and begins to fill in pore spaces above the barrier (Figure 9.5). The water table is the boundary between those pore spaces that are completely filled with water and those that are empty. Lake levels in the Interlachen map simply reflect the local water table level.

k) The water table is not always horizontal, but rather mimics the surface topography in a slightly subdued form. With a colored pencil, carefully draw in the water table in your profile.

l) Is the water table horizontal? If not, in which direction is it tilted?

This tilt is important in controlling the flow of groundwater in the region. Groundwater, like streamwater, flows downhill under the influence of gravity. It will therefore flow from high water-table elevations to lower ones.

m) Which way would groundwater be expected to flow in your profile? Show the direction with an arrow.

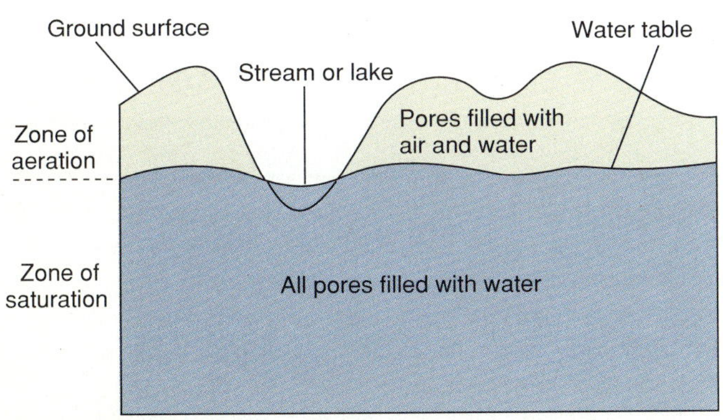

FIGURE 9.5 The water table separates the zone of saturation (where all pores are filled with water) from the zone of aeration (where pores may contain air or water). A lake or stream occurs when the water table intersects the land surface.

GROUNDWATER RESOURCES, POLLUTION, AND OTHER PROBLEMS

Groundwater flows so slowly that it does not replenish itself as rapidly as our needs might require. Go to your favorite fast-food restaurant and order a thick shake. Take the top off the container, insert a straw, and start to enjoy your dessert. As soon as you taste the shake, look at the surface of the liquid in the cup. A small depression forms around the straw because some of the underlying liquid has been removed. The same thing happens when groundwater is removed from an area, as shown in figure 9.6. The depression is called a **cone of depression** and marks the draw-down of the water table in the region. One of the problems of removing too much groundwater from an area is that large cones of depression form, making local wells run dry. Exercise 9.2 examines some of the very real problems that affect areas when groundwater supplies cannot meet the needs of all potential users.

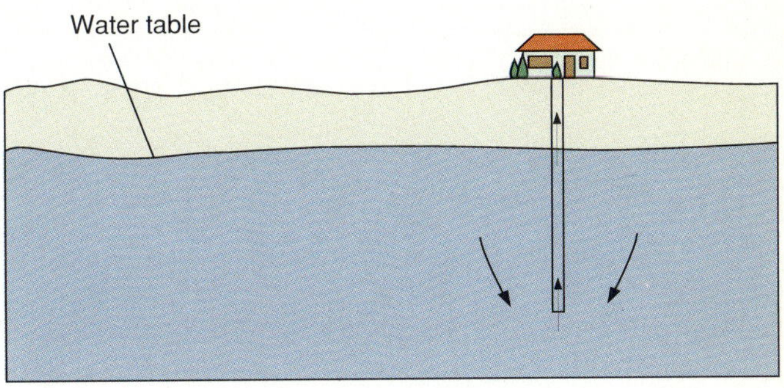

a) New home is built with a well for household water. Arrows show groundwater flow when pumping begins.

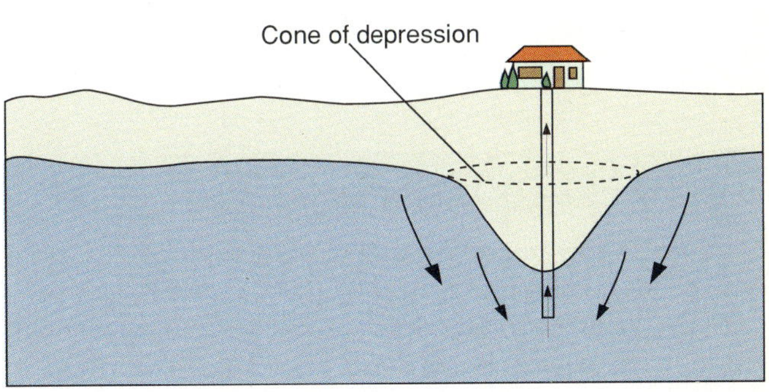

b) Gravity causes groundwater to flow downward to replace water pumped from below the water table. This lowers the water table in the vicinity of the well, forming the cone of depression. Continued pumping removes more groundwater, enlarging the cone of depression.

FIGURE 9.6 Development of a cone of depression.

EXERCISE 9.2 FARMER JONES VS. GRANDMA'S SOUP COMPANY

Figure 9.7 is a map showing an area in a midwestern state, which will remain nameless in order to protect the privacy of the individuals involved. Several farmers have been living in the area for years, each depending on groundwater supplies for drinking and irrigation. The local town selectmen succeeded in bringing two new industries into the area: Grandma's Soup Company and the Acme Sand and Gravel quarry. Trouble set in when Farmer Jones (see location on map) discovered that his well had run dry during the rainy season. Having passed Geology 100 at his state university a few years back, Farmer Jones suspected that the loss of water was due to increased water use by the new industries. As he was closer to Grandma's Soup Company than to the gravel pit, he immediately hired a lawyer and sued Grandma to stop her from pumping "his" water. Grandma argued that she had nothing to do with his problem. The judge, who never took Geology 100 anywhere, has appointed **you** as an expert to resolve the problem.

Your assignment includes the following: (1) Find out why Farmer Jones has lost his water, and (2) figure out how to keep everyone happy! (After all, Jones and his friends enjoy the tax advantage of having Grandma's factory there.)

So far, you've accumulated a lot of data. A quick survey reveals that the ground surface is as flat as a pancake. Table 9.1a shows the depth to the water table in wells throughout the area **prior to the opening of the two new companies.** Table 9.1b shows the depth to water table **after the two began pumping.** Analysis of activities at the two companies shows that Grandma uses water as a major ingredient in her soup; the Acme quarry doesn't need the water, but must pump it out of the pit so that its bulldozers can function.

Now analyze the data and make a formal recommendation to the judge. (HINT: Use the contouring skills that you acquired in chapter 6 to make a topographic map of the water table before the companies began to operate; then use tracing paper to make another contour map showing the water table after operations began.)

EXERCISE 9.3 GROUNDWATER CONTAMINATION

Quantity is one important aspect of groundwater as a resource, but **quality** is equally important. Groundwater polluted by commercial activities or by leakage from residential septic tanks may be unsuitable for drinking. This exercise deals with a common groundwater pollution problem.

Mr. and Mrs. Smith live in a suburban development in a western state. Recently, they have noticed that their drinking water has begun to taste foul, and after they wash dishes, their kitchen smells like gasoline. One evening, they watched a television newscaster describing the results of leaks from underground tanks at neighborhood gasoline stations, and realized that this might be their problem. Figure 9.8 shows their neighborhood, including two local stations that have been in business for years.

Once again, you are hired to figure out what is going on. Using state-of-the-art detection instruments, you ana-

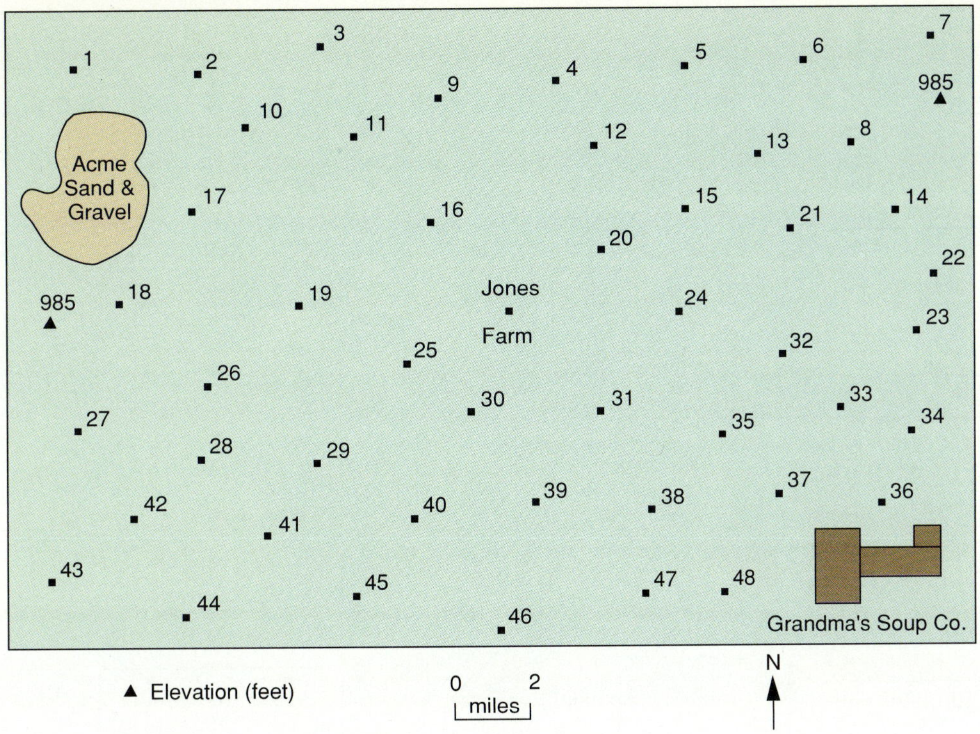

FIGURE 9.7 Farmer Jones vs. Grandma's Soup Company location map.

TABLE 9.1 DEPTH TO WATER TABLE NEAR THE JONES FARM

a) BEFORE COMPANIES BEGAN PUMPING				b) AFTER COMPANIES HAD BEEN PUMPING FOR SIX MONTHS			
Farm	Depth to water table (feet)	Farm	Depth to water table (feet)	Farm	Depth to water table (feet)	Farm	Depth to water table (feet)
Jones	32	25	40	Jones	146	25	103
1	78	26	52	1	132	26	118
2	68	27	57	2	118	27	124
3	62	28	50	3	97	28	111
4	39	29	46	4	81	29	102
5	40	30	30	5	72	30	97
6	48	31	49	6	82	31	82
7	60	32	61	7	80	32	99
8	57	33	70	8	107	33	108
9	51	34	77	9	123	34	118
10	63	35	62	10	113	35	100
11	56	36	81	11	92	36	120
12	27	37	71	12	71	37	111
13	52	38	59	13	89	38	98
14	64	39	50	14	90	39	79
15	47	40	44	15	107	40	93
16	49	41	43	16	128	41	102
17	62	42	51	17	133	42	116
18	64	43	58	18	116	43	118
19	53	44	42	19	89	44	106
20	41	45	48	20	88	45	91
21	59	46	59	21	100	46	80
22	70	47	64	22	107	47	99
23	71	48	74	23	84	48	117
24	51			24			

FIGURE 9.8 Polluted water at the Smith's; location map.

lyze the amount of a particular group of chemicals called *semi-volatile compounds* present in the soil in the Smith's neighborhood (Table 9.2). These compounds are present in most petroleum products and, because they are easily detected, are a good measure of soil and groundwater contamination. Concentrations of >50 parts per million (ppm) of these compounds are considered dangerous.

Now, answer the following questions:

1) Does either gas station have a leakage problem?

2) In which direction does local groundwater flow? (If there is a leak, and gasoline is moving along with the groundwater, which homes will be the next to be affected?)

3) What additional problems might there be in the area in the future?

(HINT: Make a contour map of the level of contamination in the soil and think about the results. Using a red pencil, lightly shade in all areas with higher than 50 ppm concentration of the semivolatile compounds. Use orange to identify areas with concentrations between 30 and 49 ppm; yellow for areas between 10 and 29 ppm. This is not a contour map that shows elevations, but the rules are much the same.)

Table 9.2 Concentrations of Semi-Volatile Compounds in Soils in the Smith Neighborhood

SITE	CONCENTRATION (ppm)	SITE	CONCENTRATION (ppm)	SITE	CONCENTRATION (ppm)	SITE	CONCENTRATION (ppm)
Smith	53	41	4	82	16	123	7
1	n.d.	42	2	83	7	124	5
2	n.d.	43	n.d.	84	12	125	n.d.
3	3	44	1	85	7	126	1
4	17	45	2	86	26	127	2
5	23	46	26	87	17	128	11
6	21	47	65	88	28	129	13
7	5	48	90	89	32	130	12
8	5	49	86	90	37	131	8
9	n.d.	50	82	91	26	132	2
10	1	51	48	92	11	133	2
11	n.d.	52	30	93	8	134	n.d.
12	n.d.	53	9	94	1	135	n.d.
13	1	54	6	95	4	136	n.d.
14	1	55	3	96	7	137	1
15	5	56	2	97	8	138	1
16	20	57	1	98	8	139	n.d.
17	18	58	3	99	6	140	1
18	56	59	10	100	25	141	1
19	53	60	12	101	44	142	2
20	61	61	8	102	49	143	6
21	2	62	20	103	n.d.	144	11
22	n.d.	63	40	104	2	145	22
23	1	64	39	105	4	146	13
24	n.d.	65	55	106	6	147	n.d.
25	n.d.	66	62	107	6	148	1
26	1	67	65	108	7	149	n.d.
27	1	68	57	109	11	150	4
28	2	69	68	110	28	151	24
29	2	70	58	111	40	152	24
30	4	71	23	112	32	153	10
31	46	72	2	113	24	154	4
32	62	73	n.d.	114	n.d.	155	2
33	65	74	1	115	4	156	n.d.
34	74	75	8	116	5	157	n.d.
35	63	76	51	117	25	158	1
36	63	77	34	118	7	159	1
37	68	78	30	119	6	160	n.d.
38	50	79	10	120	7	161	n.d.
39	31	80	23	121	7	162	1
40	10	81	25	122	11		

(Concentrations given in parts per million [ppm])
n.d. = none detected

CHAPTER 10

ARID LANDSCAPES AND THE WORK OF THE WIND

PURPOSE

To learn how the differences between processes in arid and humid regions result in different landscapes; to recognize features produced by wind erosion and deposition.

MATERIALS NEEDED

1) Colored pencils.
2) Magnifying glass or hand lens to help read close-spaced contour lines.

INTRODUCTION

Arid regions receive less than 25 cm (10 inches) of rain per year, and thus have little water for chemical weathering; stream, glacier, or groundwater erosion; or for the survival of plants and animals. Because we are conditioned by movies and television coverage of recent world events, the word "arid" brings to mind broad deserts filled with thick sheets of hot sand, large elongate sand dunes, and usually a camel or two nibbling grass near an oasis. Some arid regions, such as the Arabian, Sahara, and Mojave deserts, actually look like that (minus the camels in the Mojave), but others are rocky, frigid places, and yet others occur along the shores of oceans where there is, apparently, plenty of water. Your textbook explains how these different kinds of desert form; our goal here is to see how the scarcity of water produces landscapes that differ from those formed by streams, glaciers, and groundwater in humid regions.

A common misconception about deserts is that wind is the chief agent of erosion. Wind may play a more important erosional role in arid regions than in humid ones, but in most instances, it is still streams that do most of the erosion and deposition. In humid regions, water in the soil acts as a weak cohesive force and prevents wind from picking up small sand grains and moving them. In deserts, the soil water is absent, and the wind is able to pick up and move sand-sized grains.

EXERCISE 10.1 THE UNIQUENESS OF ARID LANDSCAPES

Look at figure 10.1, a photograph of an arid landscape in Canyonlands National Park, Utah. Compare this landscape with those of the humid areas shown in figure 7.1a and b.

---

a) How is the Canyonlands landscape different from those of the Yellowstone and Cascapedia river valleys? (HINT: Compare number of rivers, amount of vegetation, steepness of slopes, etc. Then analyze what is missing and what is the same.)

---

b) What geologic processes would you expect to be different in arid and humid areas? (To help answer this question, fill in the following checklist with the extent and nature of the processes indicated.)

	ARID REGIONS	HUMID REGIONS
1. Physical weathering		
2. Chemical weathering		
3. Soil formation		
4. Mass wasting		
5. Stream erosion		
6. Groundwater activity		

---

c) What evidence is there that stream erosion is an important part of the landscape-forming process in this arid area?

---

d) Why does the arid landscape look so different from the humid ones?

Because of the scarcity of water, weathering and erosion take much longer in arid regions than in humid ones. As a result, features formed by the **constructional processes** mentioned at the beginning of Section II survive much longer than in humid regions, and weathering accentuates slight differences in rock types. As an example, look at figure 10.2, a photograph of part of the Grand Canyon of the Colorado River in Arizona.

Note that the cliff faces do not all have the same degree of steepness; some are gentler and others steeper. Steep slopes represent more resistant rock types; gentler slopes, those that are more easily eroded or more subject to mass wasting. Abrupt changes in slope indicate contacts between different rock types.

FIGURE 10.1 Canyonlands National Park, Utah, showing features typical of stream erosion in an arid environment.

FIGURE 10.2 Part of Grand Canyon National Park, Arizona. The steepness of different cliff slopes defines rock layers with different erosion levels. Steep slopes form inconsistent layers and gentle slopes have easily erodible layers.

---

a) Trace the top of one resistant bed on the photograph with a colored pencil; then trace the bottom contact of that bed. Color that layer of rock.

---

b) Do the same with layers above and below.

It is much easier to trace individual layers in an arid region than in a humid area, because creep, slump, and development of soil in humid areas mask the differences in rock type to a great extent (you cannot see them, for example, in figure 7.1a or b). Geological mapping is thus often easier in arid regions than in humid ones, because different rock types can be so readily identified from a topographic map or aerial photograph, and because there is little soil and vegetation to obscure bedrock relations.

---

c) Explain why creep, slump, and soil formation have not masked the differences in rock type here.

Evolution of Landscapes in Arid Regions

Constructional landforms such as uplifted fault blocks or volcanoes are gradually eroded, even in arid regions, although erosion is slower and the erosional stages are quite different from those of humid regions (as seen in figure 7.9 and studied in Exercise 7.4). Figure 10.3 describes the stages of erosion in an arid region. Compare these with the stages in humid regions (Figure 7.9).

The area shown has been affected by block faulting, much like that in the southwestern United States. Large areas have been uplifted and others downdropped by these faults, forming elongate uplands and lowlands (Figure 10.3a). In early stages of arid erosion, most of the uplands area is preserved, and the faults themselves may be visible as unusually straight topographic lineaments. What little stream erosion occurs does so slowly, carving into the uplifted blocks. This eventually results in a rugged, mountainous topography characteristic of the intermediate stage of arid landscape development (Figure 10.3b). Sediment eroded from the mountains is deposited by the streams as alluvial fans at the bases of the mountains, filling the lowlands from both sides. Eventually, the alluvial fans merge into broad features, called bajadas (ba-ha'-das), that slope down into the lowlands from the eroded highland blocks. Streams that enter the lowland have no way to get out, and their waters form lakes that may partially fill the depressions. These are called **playa lakes** or simply **playas.**

Filling in the lowlands and lowering the highlands causes stream gradients to decrease. The streams begin to meander and carve a flat, sloping surface into the bedrock, which is called a **pediment** (Figure 10.3b). Continued deposition buries the pediment and fills in the lowland completely. In the late stage of the arid erosional cycle (Figure 10.3c), thick blankets of sediment mask all but a few remnants (called **inselbergs**—island mountains) of the original highlands.

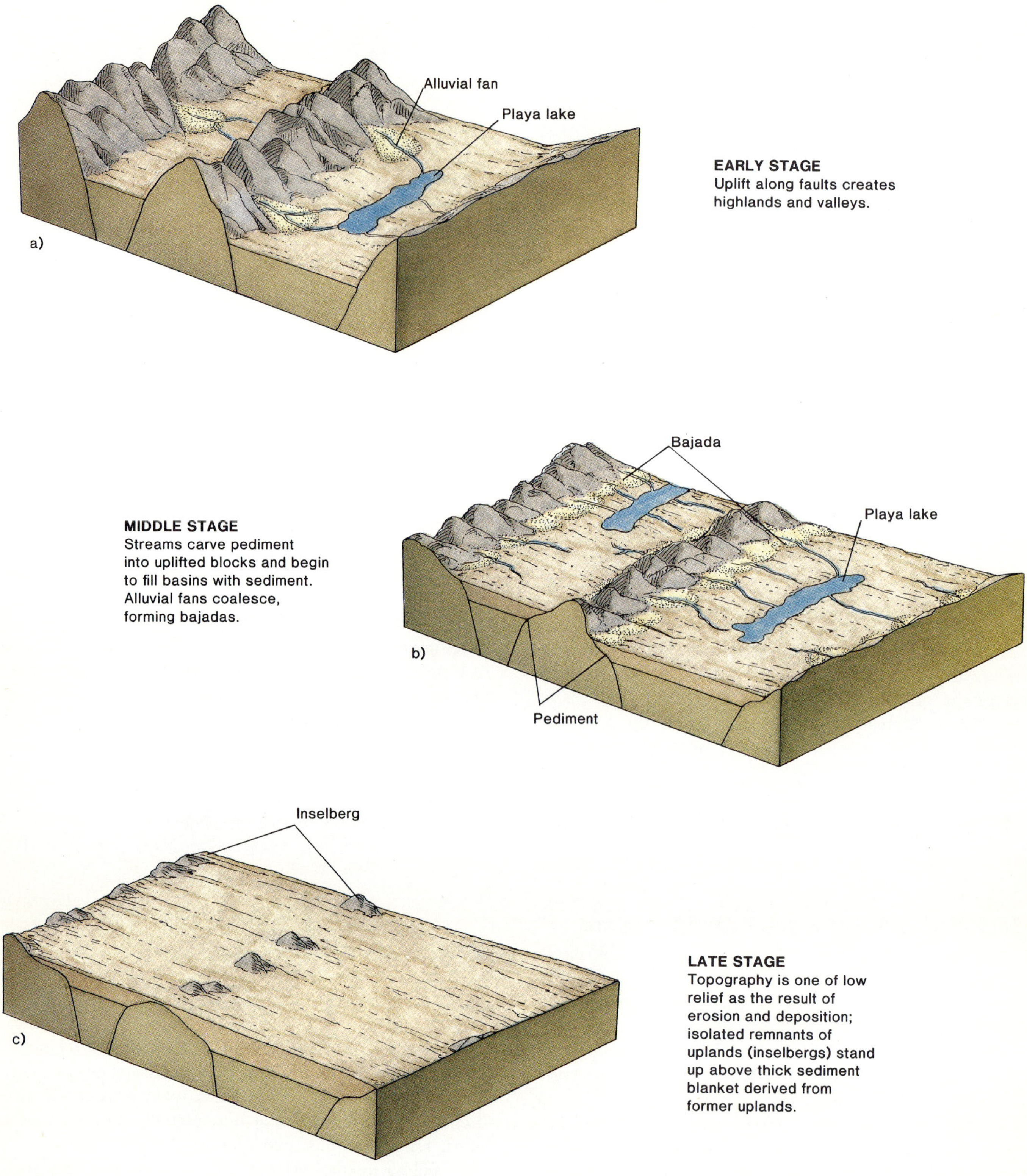

FIGURE 10.3 Stages of erosion in arid regions.

EXERCISE 10.2 EXAMINATION OF ARID LANDSCAPES

Examine the portion of the Furnace Creek, California/ Nevada, Quadrangle shown in figure 10.4. The area is part of Death Valley National Monument and is one of the most arid places in North America.

a) Note that the streams are shown with dashed blue lines instead of solid lines. Suggest an explanation.

b) What kind of lake is shown in the map? Why is it shown in a dashed pattern?

c) What evidence is there that you probably would not want to have a refreshing drink from this lake, even though you are in a desert? Suggest a reason for the non-drinkability of the water.

d) Why can the water that flows into this lake never escape into the ocean? (HINT: What is the elevation of the land around the lake?)

e) With a colored pencil, draw lines to indicate the locations of faults that separate uplifted from downdropped blocks. How many such blocks are there? Which ones are easier to locate than others? Explain why.

f) Draw a heavy line with a colored pencil to outline an alluvial fan on the west side of the map, and lightly color the fan. Do the same with others in that area. Now try to do the same with fans in the northeastern part of the map.

What is the difference between the two sets of fans?

g) Sketch or construct a topographic profile from the northeast corner of the map area to the southwest. Compare this profile with figure 10.3.

What is the difference between Death Valley and the valley separating the Funeral and Black Mountains?

Can you detect any signs of the original fault surfaces? If so, where?

Locate inselbergs, bajadas, and pediments if you can.

What stage (or stages) of the arid erosional cycle are represented on this map? What is your evidence?

Now examine the part of the Antelope Peak, Arizona, Quadrangle shown in figure 10.5.

FIGURE 10.4 Part of the Furnace Creek, Nevada Quadrangle. Scale: 1:62,500. Contour interval: 40′.

FIGURE 10.5 Antelope Peak, Arizona. Scale: 1:62,500. contour interval: 25′.

h) Compare this stage in the arid erosional cycle with that of the Furnace Creek area. (HINT: locate faults, inselbergs, bajadas, and pediments if you can, and compare with results in (g) on page 131.)

The Work of the Wind

Wind is an active agent of erosion in arid regions for reasons mentioned earlier, but it is also important in humid areas, as long as two important conditions are met: (1) there must be an ample supply of fine-grained (sand-sized) sediment, and (2) there must be little if any vegetation. Plant roots, stems, and leaves commonly trap sand grains and prevent the wind from picking them up and moving them. Two types of humid areas commonly display features of wind erosion: shorelines where large amounts of loose sand are available, and regions adjacent to glacial outwash plains. After a glacier retreats, wind blowing across the outwash plain can pick up very fine grains and transport them. The resulting deposits are referred to as **loess.**

Wind erodes bedrock by abrasion, in much the same way that a stream does. Sand grains carried by the wind collide with outcrops of bedrock, literally sandblasting the outcrops away. Swirling eddies of wind frequently scour loose sand from the ground, carving a circular pit, aptly called a **blowout.**

Landforms deposited by the wind are called **sand dunes,** and are nothing more than giant ripplemarks. With a steady wind and a large supply of sand, dunes of several types form and then migrate across the land surface until they reach an obstruction or are stabilized by plant roots and stems (Figure 10.6). There are several kinds of dunes, each forming under special conditions of wind velocity, direction, and the amount of sand available (Figure 10.7).

Topography dominated by sand dunes tends to be as irregular as the knob-and-kettle topography of glacial ground moraine or the sinkholes of karst regions. Indeed, the three may look similar; blowouts can often be mistaken for kettles or sinkholes. Location (well outside glaciated areas), systematic orientation and asymmetry of dunes (instead of the much less regular karst features), and size help identify an area composed of wind-produced landforms. Sand-dune topography is also temporary: individual dunes migrate, several dunes may coalesce, or a large dune may split into smaller ones, and blowouts may form literally overnight during a single storm.

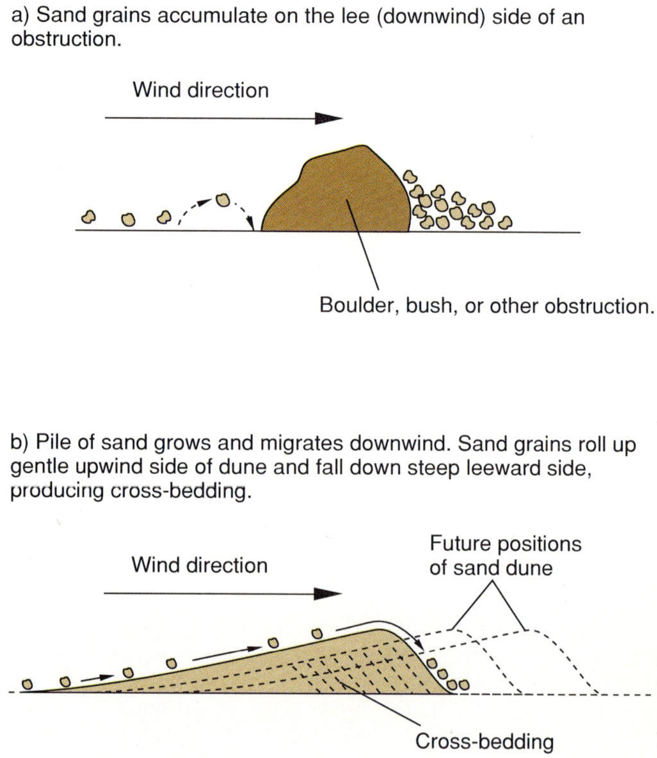

FIGURE 10.6 Formation and migration of sand dunes.

Exercise 10.3 Landforms Produced by the Wind

Examine the part of the Moses Lake, Washington, Quadrangle shown in figure 10.8.

a) Describe the topography in the southern third of the map area in your own words.

b) How could you be certain that this topography is not an end moraine or part of a karst region?

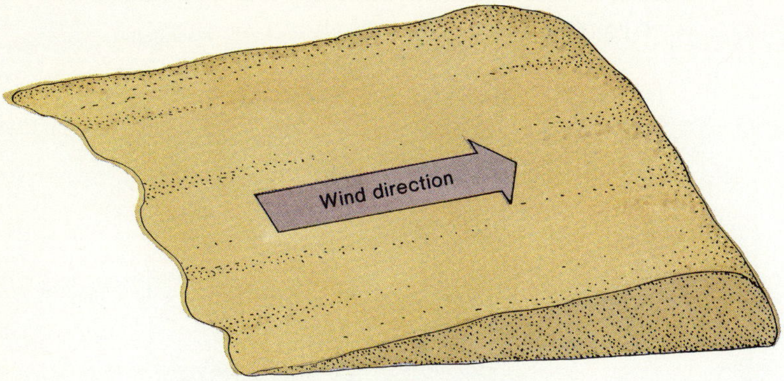

Transverse Dune: Forms at right angles to wind direction in areas of abundant sand and steady, moderate to strong winds. **Size:** 10's of feet high, a few miles long.

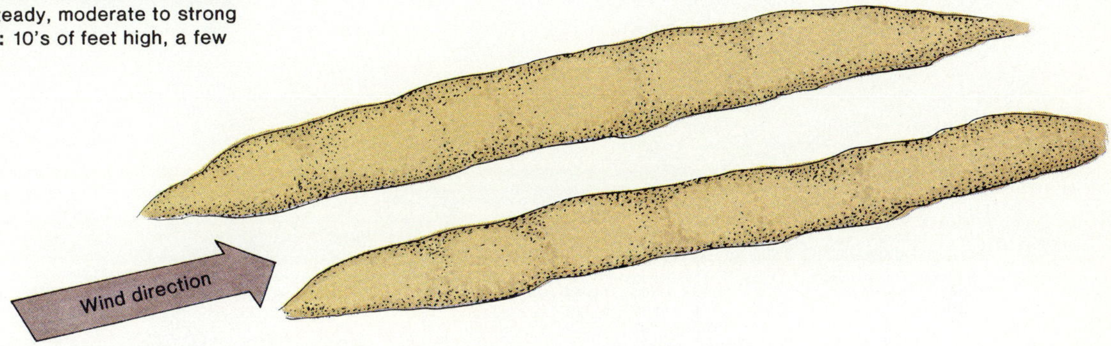

Longitudinal Dunes: Form parallel to wind direction in areas of enormous sand supply and high wind velocity (e.g., centers of large, sandy deserts such as the Sahara). **Size:** 100's of feet high; up to 100 miles long.

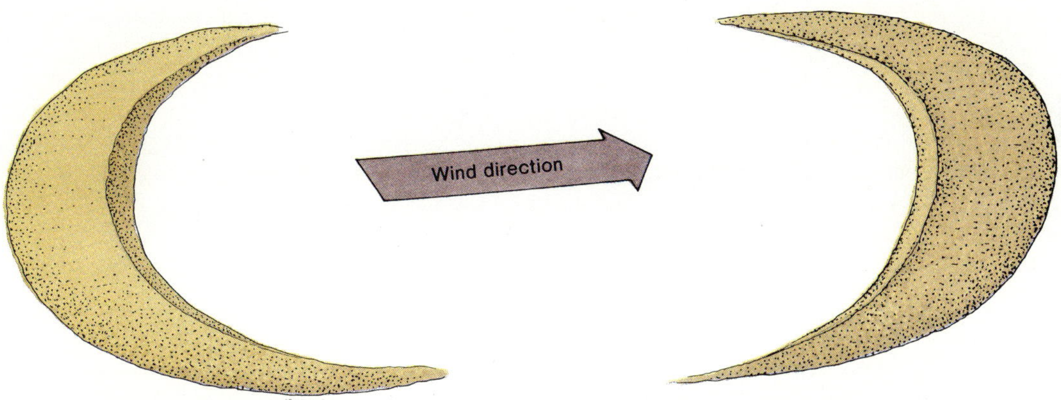

Barchan: Forms where there is a limited amount of sand and a constant, moderate wind.

Parabolic (blowout) dune: Forms where vegetation anchors the ends of a transverse dune or where a blowout removes part of a transverse dune.

FIGURE 10.7 Types of sand dunes and how they form.

FIGURE 10.8 Part of the Moses Lake, Washington Quadrangle. Scale: 1:62,500. Contour interval: 25′.

c) The shapes of the dunes in the southwest corner and southeast quarter of the map are distinctive. What kinds of dunes can these be? Which direction(s) would the wind be blowing for each type?

In order to tell which kind of dune these are, and to determine the dominant wind direction, a little detective work is necessary. First, look again at figure 10.7 to see the relations between transverse dunes and the two types possibly present here.

d) Several linear dunes are present in the area south of Camas Mounds School. Are these transverse or longitudinal dunes? Cite your evidence.

e) Sketch an east-west profile across the elongate dune that is approximately 0.5 miles west of the 1107-foot bench mark in the NE¼ of Section 14, T18N, R27E. Which way is the wind blowing that formed this dune?

f) Look at the two smaller dunes immediately east of this one. Knowing what you now do about the wind direction in the area, how did these form? Which kind of dune do these represent? Finally, which type is present in the southwest and southeast parts of the map? Cite your evidence for all these answers.

g) What is the origin of "The Potholes"?

h) Explain why lakes are present in many of these features. (HINT: This answer has nothing to do with arid regions or wind erosion.)

Examine the part of the Spade Ranch, Nebraska, Quadrangle shown in figure 10.9. This quadrangle is located in an extensive and unique topographic region, known as the Sand Hills of Nebraska, that extends for many miles beyond the area shown by this map.

a) Describe the topography of the map area in your own words. How is it different from that of the Moses Lake quadrangle?

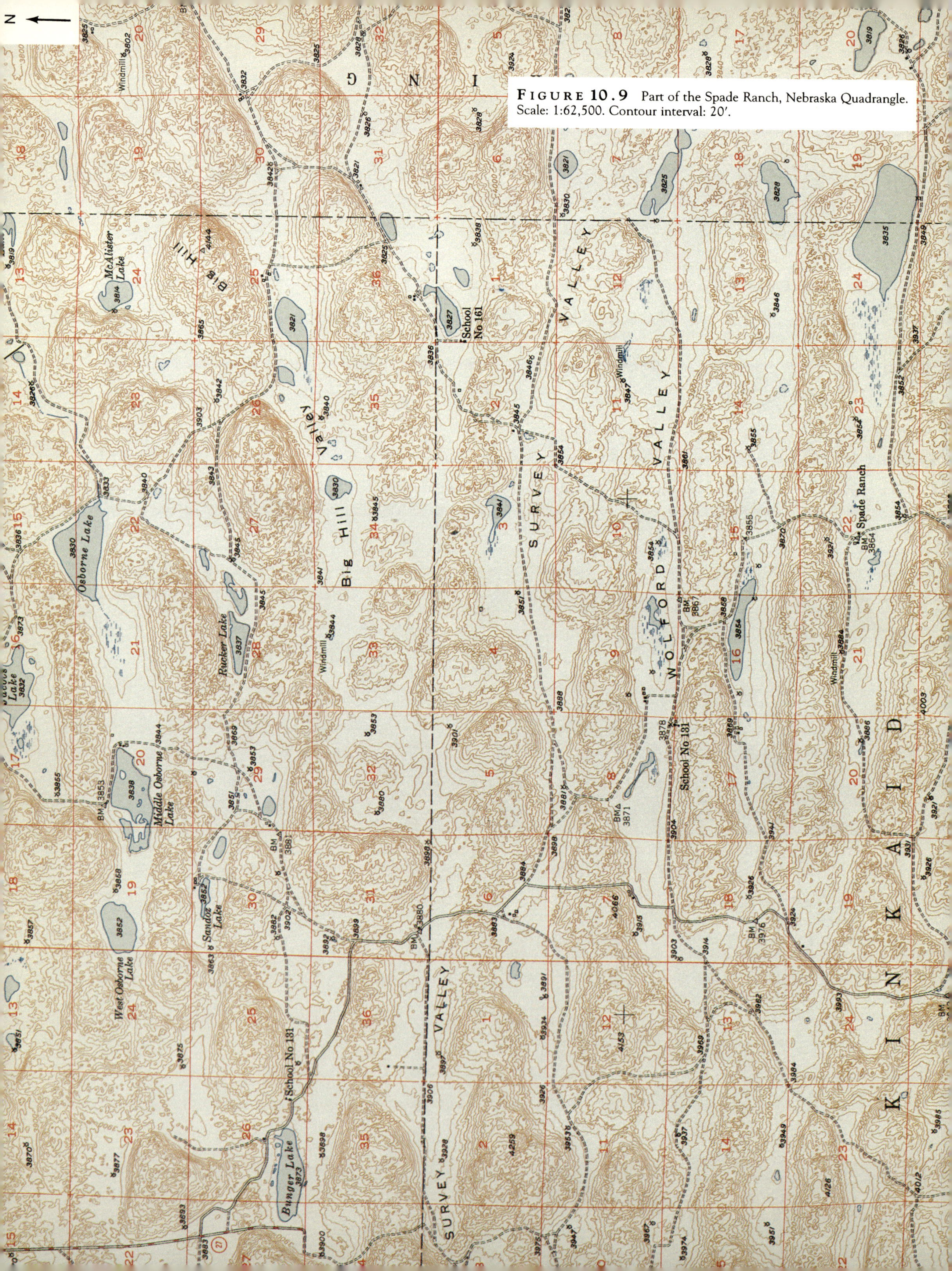

FIGURE 10.9 Part of the Spade Ranch, Nebraska Quadrangle. Scale: 1:62,500. Contour interval: 20′.

b) What evidence indicates that the landscape in this area is not the result of groundwater solution or glacial deposition?

c) What is the origin of the elongate valleys indicated by the hachured contour lines?

d) Just on general principles, describe the tilt of the water table in this area.

e) What factors can explain why this area looks so different from that of the Moses Lake quadrangle?

f) The long, linear sand dunes present here may be either transverse or longitudinal. How could you tell them apart?

g) O.K. Now look carefully at the dunes, and decide which kind is present and which way the wind is blowing. What is your evidence?

Although the Spade Ranch Quadrangle clearly shows the results of wind action, the presence of so many lakes indicates that the area is not entirely arid. This leads to some more sophisticated thinking and detective work.

a) The absence of streams in karst topography is explained by the fact that most rainwater infiltrates the ground through fractures and sinkholes. What explanation can you suggest for the absence of streams in the Spade Ranch area? (HINT: Think about what the materials present at the surface must be and what their physical properties are.)

b) Arid landscape features in a moderately humid region seem illogical. Suggest an explanation for this apparent paradox. (HINT: Earth is a dynamic planet, few of whose features remain constant with time. The observations that led to this apparent paradox and the reasoning that is associated with its resolution provide a good example of how a keen pair of eyes and a questioning mind can lead to important hypotheses about the evolution of the Earth.)

CHAPTER 11

SHORELINES: COASTAL EROSION AND DEPOSITION

PURPOSE

To learn the features that develop along different kinds of shorelines; to use these features to interpret the processes active along lakeshores and seashores.

MATERIALS NEEDED

1) Colored pencils.
2) Magnifying glass or hand lens to help read close-spaced contour lines.

Introduction

Shorelines, where the land meets the sea or a large lake, are among the most complex geologic environments on Earth. It is along shorelines that wind-driven waves crash onto the land, eroding it and redistributing the eroded sediment along the shore. Wind and water move very quickly and can cause rapid changes; the entire configuration of a coastline can change overnight after a major storm. The complexities arise because several completely different sets of processes combine to produce shoreline features:

Tectonic processes uplift or depress the land, causing sea level to change.

Sea-floor spreading and subduction enlarge or shrink ocean basins, also causing the water level to drop or rise.

The type of material of which the land is made controls the erosional and depositional features that are produced. Loose, unconsolidated sediment is easily eroded and redeposited, but a shoreline made of resistant igneous and metamorphic rocks will respond very differently to the same erosional agents.

The landforms found along a shoreline result from a balance between wave-caused shoreline erosion and deposition and the action of streams and glaciers.

The wind that causes waves to move across the surface of the ocean (or lake) may change direction suddenly and thus attack the shoreline from a completely different direction.

Vastly different **rates** of activity add to the complexity of shorelines. For example, sea-floor spreading and subduction create and eventually destroy ocean basins, but their effects on existing shorelines are much slower and less noticeable to us than those of local streams, glaciers, and wind. Most modern shorelines are still in a state of disequilibrium, not yet having achieved a balance between local stream, glacial, wind, and tectonic processes, and the results of melting the enormous Pleistocene continental ice sheets.

Although the coastal environment is complex, it is absolutely critical that we understand it, because such a large portion of the world's population lives on shorelines or immediately adjacent to them. A little common sense, coupled with a few bits of geological information, will enable you to figure out most shoreline processes.

The following exercises are designed to first introduce you in more detail to some of the variables listed earlier, and then to let you apply your knowledge to analyze the origin and evolution of selected shoreline segments.

EXERCISE 11.1 FACTORS IN SHORELINE DEVELOPMENT

Waves

Wind blowing across the surface of the open ocean (or of a large lake) generates waves that crash onshore and cause most shoreline erosion and deposition. Figure 11.1 shows waves in the Pacific Ocean coming onshore in California (a) and Hawaii (b).

a) Compare and contrast the two sets of waves.

b) Using a colored pencil, draw arrows to indicate the wind direction responsible for the waves in each area.

c) In which area is the wind stronger? Explain.

d) Based on your answer to (c), where would you expect wave erosion to be most effective? What factors other than wind velocity must be considered?

Coastal Materials

Figure 11.2 shows that shorelines can be made of very different kinds of materials, ranging from loose sediment (a) to walls of solid bedrock (b). These photographs show the two extremes; intermediate kinds of material include weakly lithified sedimentary rock or deeply weathered bedrock.

e) Describe ways in which waves can erode **both** types of material shown here.

f) Which would be more easily eroded? Why?

FIGURE 11.1 Waves.

FIGURE 11.2 Shoreline materials. a) Gentle beach slope made of loose sand in Hawaii. b) Steep, rocky cliff in Maine showing bedrock exposures, layering.

---

g) In which area would you expect shoreline changes to be most rapid? Explain.

---

h) In which area would you expect the most extensive shoreline **depositional** features? Explain.

Where there is an abundant supply of sand along a shoreline, a distinctive series of depositional features forms when waves and resulting longshore currents shift the loose sediment from place to place. Figure 11.3 shows some of the common depositional and erosional features found along such a coast.

Progressive Erosion Along Shorelines

The ideal end product of shoreline erosion and deposition is a relatively straight coastline. Wave refraction focuses kinetic energy on promontories that extend into the sea, and disperses energy broadly in the embayments. Promontories thus tend to be eroded rapidly, while embayments are filled with the resulting sediment. The result is a relatively straight coast with few headlands or bays. The complex interaction of processes along shorelines, however, makes sure this ideal is rarely achieved.

Erosion occurs most readily at the weakest parts of the promontories, and these are excavated first (Figure 11.4). Eventually, the remainder of the headland is also carved away by the waves (Figure 11.5).

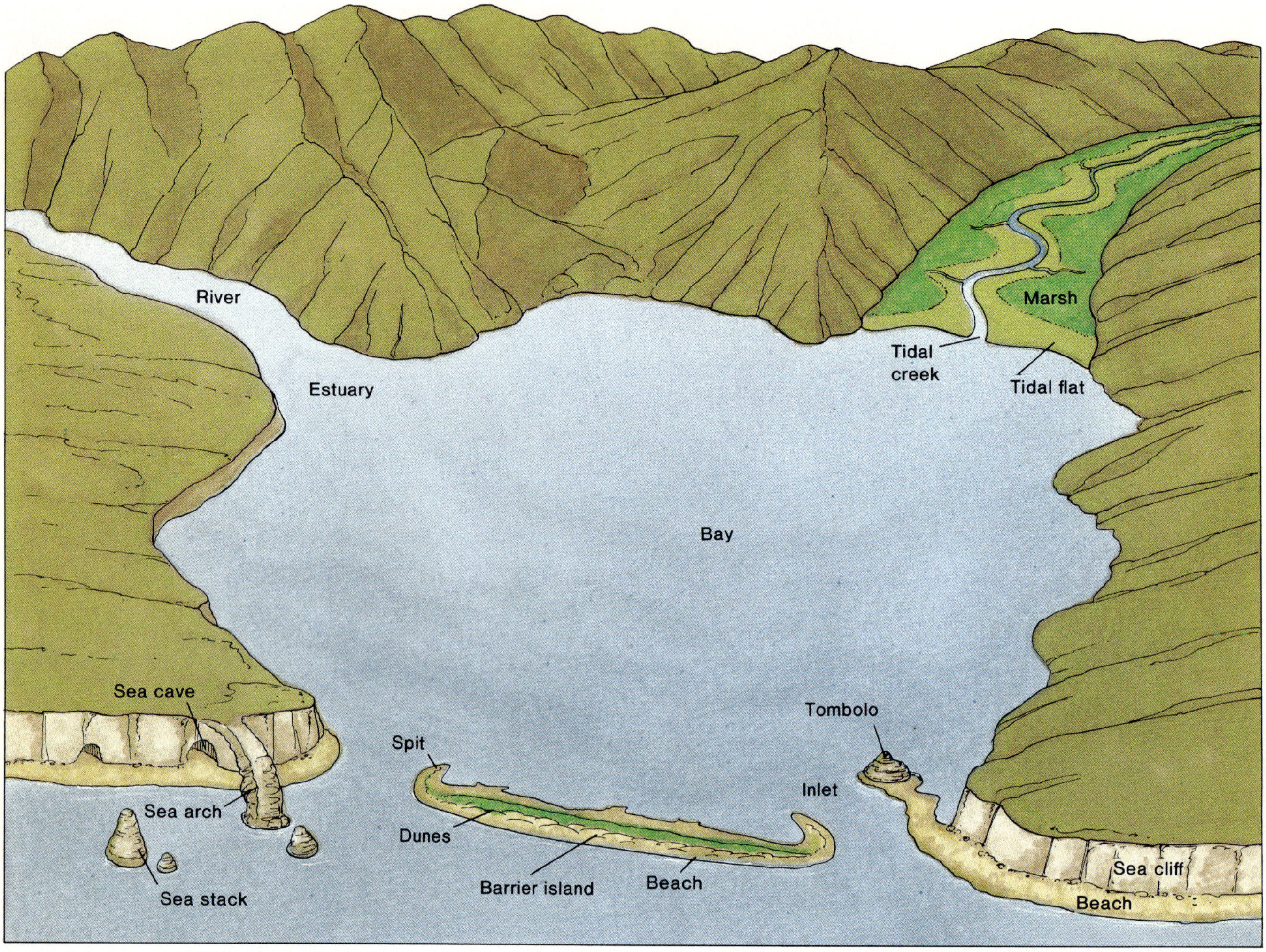

FIGURE 11.3 Typical coastal landforms.

---

i) The feature shown in Figure 11.4a is called a **sea arch,** for obvious reasons. This one is carved out of lava on the island of Hawaii. Suggest a reason for the more rapid erosion of the "archway" as compared to the pillar.

---

j) Alone, water moving against bedrock would accomplish little in the way of erosion. How can waves carve their way into solid bedrock? (HINT: remember how streams and glaciers erode.)

---

k) Since the waves can erode only the bottom part of the cliff, suggest a mechanism by which the entire cliff face can be eroded back into the land.

Eventually, mass wasting will cause the top of the arch to collapse, isolating the pillar offshore as a **sea stack.** Figure 11.4b shows Percé Rock, a large sea stack off the easternmost point of the Gaspé Peninsula of Quebec. At one time, it was part of a continuous bedrock mass extending to the place from which the photograph was taken.

Once local irregularities have been removed, large-scale wave action carves a **wave-cut cliff** into bedrock along the shore (Figure 11.1b) and eventually eats away at the cliff, causing the shoreline to recede, and leaving behind a **wave-cut bench** (Figure 11.5).

a)

b)

FIGURE 11.4 Isolation and removal of bedrock along a shore. a) Sea arch on the south shore of the island of Hawaii; arch is developed in basaltic lava. b) Percé Rock, eastern tip of Gaspé Peninsula, Quebec.

FIGURE 11.5 Photograph showing a wave-cut bench in California.

Emergence vs. Submergence

If sea level along a shoreline changes, the action of the waves will affect points that are either higher or lower on the land surface. When sea level drops or the land surface rises, the effect is to lower the level at which wave action operates. Such shorelines are said to be **emergent,** since they seem to be rising up from the ocean with time. Where sea level rises or the land subsides, shorelines are said to be **submergent.**

---

l) Suggest a way in which global climate changes could produce global emergence or submergence of the present coastlines.

---

m) If the Atlantic Ocean were to continue to grow by sea-floor spreading, what effect would this have on the Atlantic coast of North America? Would the coast become emergent or submergent? Explain.

Evidence for emergence or submergence frequently is visible in shoreline features. A series of maps made over a period of decades or centuries would show these changes most clearly. Submergence would simply "drown" the coastline; the sea would invade the land, and successive contour lines would eventually become the zero elevation—the shoreline. Emergence is somewhat easier to detect.

With a little thought, it is possible to figure out whether sea level has gone down or the land has been uplifted tectonically. Since the last major glacial advance during the Pleistocene Epoch, what has happened to the former continental ice sheets? What effect has that had on worldwide sea level?

n) Carefully examine figure 11.5 again, this time for evidence of emergence. (HINT: Apply the Principle of Uniformitarianism. Look at the wave-cut bench now being produced and figure out where older benches should be on an emergent coastline.) What evidence do you see?

o) With your colored pencils, indicate the amount of uplift (or sea-level lowering) that has occurred here.

p) What do these thoughts suggest about the cause of emergence here on the California coast?

Exercise 11.2 Analysis of Shoreline Features

By putting together all the factors discussed earlier, you should be able to analyze any shoreline and construct hypotheses about its composition and evolution.

Examine figure 11.6, a portion of the Boothbay, Maine, Quadrangle.

a) Is this shoreline composed mostly of loose, unconsolidated material or of solid bedrock? Give evidence for your hypothesis. (HINTS: a. Compare this map with figures 11.2 and 11.3b. This is a glaciated area; examine the north-south profile of the hill just north of the town of Linekin.)

b) Asterisks just south of each of the major promontories indicate isolated masses of rock, sometimes labeled "ledges." In view of your answer to the previous question, suggest a possible origin for these features.

c) Is this shoreline emergent or submergent? Cite evidence for your hypothesis.

FIGURE 11.6 Boothbay, Maine. Scale: 1:62,500. Contour interval: 20′.

Now examine figure 11.7, a portion of the Oceanside, California, Quadrangle.

---

d) Describe this shoreline in your own words.

---

e) In what way(s) would you expect the evolution of this shoreline to have been different from that of the Boothbay area?

---

f) Several small streams and the San Luis Rey River (just north of the town of Oceanside) flow westward from the mountains toward the Pacific Ocean, but never flow into the sea! Suggest an explanation based on shoreline erosional and depositional processes.

---

g) Suggest an origin for the Batiquitos Lagoon (in the southern part of the map).

---

h) Sketch or construct a profile from the shoreline at Carlsbad to Mt. Kelly. What evidence does this profile provide concerning the emergence or submergence of the coastline?

The next two shoreline examples illustrate the rapid changes that characterize some areas. First, compare figures 11.8 and 11.9, topographic maps of the Provincetown, Massachusetts, Quadrangle made in 1889 (Figure 11.8) and 1972 (Figure 11.9).

---

i) Of what type of material is this shoreline composed? Be as specific as you can. Compare with the Boothbay and Oceanside areas.

---

j) Describe the changes that have taken place in the 83 years that separate the two maps. (HINT: Use one colored pencil to show the depositional changes on the old map, a different one to show the erosional changes.)

---

k) With another pencil, draw arrows to indicate the longshore drift directions for this area.

FIGURE 11.7 Oceanside, California. Scale: 1:62,500. Contour interval: 25′.

FIGURE 11.8 Part of the Provincetown, Massachusetts Quadrangle. Date: 1889. Scale: 1:62,500. Contour interval: 40′.

FIGURE 11.9 Part of the Provincetown, Massachusetts Quadrangle. Date: 1972. Scale: 1:38,340. Contour interval: 10′.

l) A large "dike" is shown on the 1972 map extending southward from a point south of Telegraph Hill toward Wood End on Long Point. What effect has this structure had on deposition and erosion in the western part of Provincetown Harbor?

m) Blue contour lines offshore show the slope of the sea floor on the 1972 map. [Note that the contour interval is different from that used on land. Suggest a reason for the 6-foot interval.]

Sketch a profile at right angles to the shoreline from a point 2,000 feet north of Peaked Hill Bar (in the Atlantic Ocean north of Cape Cod) to a point in the center of Provincetown Harbor. Describe the differences between the Atlantic shore and the Harbor shore, and suggest an explanation for those differences.

Use a colored pencil to show how Cape Cod would change if sea level were to drop by 15 feet.

n) Compare these maps with figure 11.3. Name the depositional features present in this area. What kind of feature is the part of Cape Cod where Provincetown is located?

Compare figures 11.10 and 11.11, parts of the Kingston, Rhode Island, Quadrangle, made in 1957 and 1975, respectively. Locate this quadrangle on the glacial map in Chapter 8.

o) What **glacial** depositional features are present in this map area?

p) Compare the shoreline materials of this area with those of the preceding maps.

q) Use a colored pencil to trace the direction(s) of longshore drift.

r) Compare the appearance of Trustom Pond and Card Ponds on the two maps. What has happened, and why?

FIGURE 11.10 Part of the Kingston, Rhode Island Quadrangle. Date: 1957. Scale: 1:24,000. Contour interval: 10′.

FIGURE 11.11 Part of the Kingston, Rhode Island Quadrangle. Date: 1975. Scale: 1:24,000. Contour interval: 10′.

s) Suggest the eventual fate of these ponds.

t) Name and locate the shoreline depositional features present on this map.

Figure 11.12 is a change of pace, a shoreline of one of the Great Lakes rather than an ocean shoreline. By now, you should need no hints to figure out the evolution of this shoreline, but the following questions should help to focus your thinking.

u) Suggest an origin for the several ridges that parallel the shore of Lake Erie (North Ridge, Sugar Ridge, Chestnut Ridge, etc.).

v) Describe the evolution of the shoreline of Lake Erie. A sketched profile from Butternut Ridge to the lake would be very enlightening.

Figure 11.13 is a part of the Rehoboth, Delaware, Quadrangle. Without any further help, describe the nature and evolution of this shoreline and its erosional and depositional features.

EXERCISE 11.3 THE GREENHOUSE EFFECT AND ITS IMPACT ON SHORELINES

The validity of the Greenhouse Effect (progressive warming of the Earth caused by the addition of carbon dioxide to the atmosphere from the burning of fossil fuels) is much debated. If Earth's surface warmed significantly, the continental ice sheets of Antarctica and Greenland would melt, at least partially, and sea level would rise. No agreement has been reached on exactly how much sea level would rise, but there is no doubt that it would.

With a brightly colored pencil, show, by tracing the new shoreline that would form, the effect of a 100-foot rise in sea level on each of the shoreline maps you studied. Use a different pencil to show the effect of a 200-foot rise.

a) Which shorelines would be affected more than the others?

b) What do these shorelines have in common?

c) What general statement can you make about the survival of our current coastal cities based on your answer to questions a and b?

FIGURE 11.12 Oberlin, Ohio. Scale: 1:62,500. Contour interval: 10′.

FIGURE 11.13 Part of the Rehoboth, Delaware Quadrangle. Scale: 1:62,500. Contour interval: 10′.

Internal Earth Processes Revealed by Surface Features

Section 3

In Sections 1 and 2, you learned the techniques by which geologists interpret different aspects of Earth history from minerals, rocks, and landforms. Section 1 focused on how to reconstruct the nature of rock-forming processes based on the mineralogy and texture of a given rock. Section 2 showed you how to identify the agents responsible for creating landscape features, and ways to interpret the erosional and depositional history of an area. In Section 3, we turn to processes that operate deep beneath the surface—the plate tectonic interactions that build large constructional landforms such as mountains and ocean basins.

We will again be looking for clues at the Earth's surface, features that will reveal the nature, intensity, and sequence of plate interactions. The task is divided into three steps corresponding to the three chapters of the Section. Chapter 12 deals with recognition of features associated with intense deformation of rocks. The first part of the chapter introduces the large-scale landscape features that indicate the presence of highly deformed rock. Geologic maps are then described—maps that show the distribution of rocks of different types and ages, as well as topography. These maps are then used to analyze the deformational history of the United States.

Chapter 13 shows how the sequence of geologic events in an area can be deciphered, and is designed to further sharpen your skills in geological observation and interpretation. Geologic maps, topographic maps, and photographs provide the clues, and you are asked to describe the detailed geologic history of the areas shown: their complete depositional, deformational, and erosional evolution.

Chapter 14 focuses entirely on unraveling large-scale plate motion. Analysis of selected plate boundaries will show how geologists try to establish not only that plates are moving, but also the rates of motion and of their interaction.

These new skills will complete your training as geological detectives. Couple these skills with the techniques of analysis and interpretation learned in Sections 1 and 2, and you will be ready to read much of the history of the Earth from its minerals, rocks, and landforms (Chapter 15).

CHAPTER 12

INTERPRETING GEOLOGIC STRUCTURES

PURPOSE

To show how landscape features reveal the presence of deformed rocks in an area; to interpret the deformational history of those rocks.

MATERIALS NEEDED

1) Colored pencils.
2) A fine-tipped black pen.
3) Tracing paper.
4) A protractor.

Introduction

Rocks are intensely deformed when the plates they compose collide in subduction zones or continent-continent collisions, rift apart as the result of sea-floor spreading, or grind past one another in transform faults. On the grand scale, plate interactions produce mountain systems, ocean basins, ridges, trenches, and fracture systems. More local results are **geologic structures**—the different kinds of folds and faults discussed in your textbook. Folds are caused largely by compression between converging plates; different kinds of faults arise from compression, rifting, and shear.

Erosion of folded and faulted rocks commonly produces distinctive topographic features. In this chapter you will learn new ways to analyze landforms—ways that reveal whether or not an area has been deformed and the details of that deformation. You will also be introduced to a new kind of map—**the geologic map**—which shows the distribution of rocks of different types and ages. These maps trace individual folds, revealing the scale and intensity of the folding, and show where rocks are truncated and offset by faults. Finally, you will learn how to construct **geologic sections** (the structural geologist's analogue of a topographic profile), which show how surface features can be extrapolated downward to reveal the location of particular layers of rock below the surface.

Throughout this chapter, we will concentrate on the geologic structures that result when sedimentary rocks are deformed. Do not conclude from this focus that igneous and metamorphic rocks are never deformed; it is simply that folds and faults are easiest to recognize in sedimentary rocks. Because sedimentary rocks are generally deposited in horizontal layers, a significant deviation from the horizontal can show how much tilting, folding, or faulting has gone on.

The Relation between Topography and Deformational History

Once again, observation coupled with common sense will enable you to derive most of the "rules" or clues to be used in analyzing an area for features that result from deformation. There is only one thing to memorize: **Some rocks are more resistant to weathering and erosion than others.** Everything else can be derived from this basic principle.

Exercise 12.1 Topographic Clues to Deformation

Figure 12.1 shows three **block diagrams,** which are three-dimensional drawings of rocks that have been subjected to different degrees of deformation: no deformation at all, simple tilting, and folding. The patterns that distinguish the sedimentary strata are the standard ones used on cross sections to indicate different kinds of rock. Each block diagram consists of a map view, showing what is visible at the surface, and two cross sections that represent the geologist's interpretation of the subsurface distribution of rock units.

a) In which of the three areas would you find the smallest number of different rock types at the surface? Explain.

b) In which of the three areas would you expect ridges and valleys to form as the result of differential weathering and erosion? Explain.

c) In which of the three areas would such elongate topographic features be absent? Explain.

d) Generalize: How can the shapes of topographic highs and lows reveal the presence of deformed rocks?

When streams erode an area, their paths are commonly controlled by the erodability of the rocks present. The arrangement of tributaries and master streams depends on how homogeneous or varied the rocks are. Specific arrangements called **drainage patterns** result from different states of deformation or other factors that cause differential erodability. For example, look again at figure 12.1. Assume that sandstone is the most resistant rock, and limestone the least resistant, to weathering and erosion.

e) In which area would you expect streams to have the most random orientations as they erode the landscape? Explain.

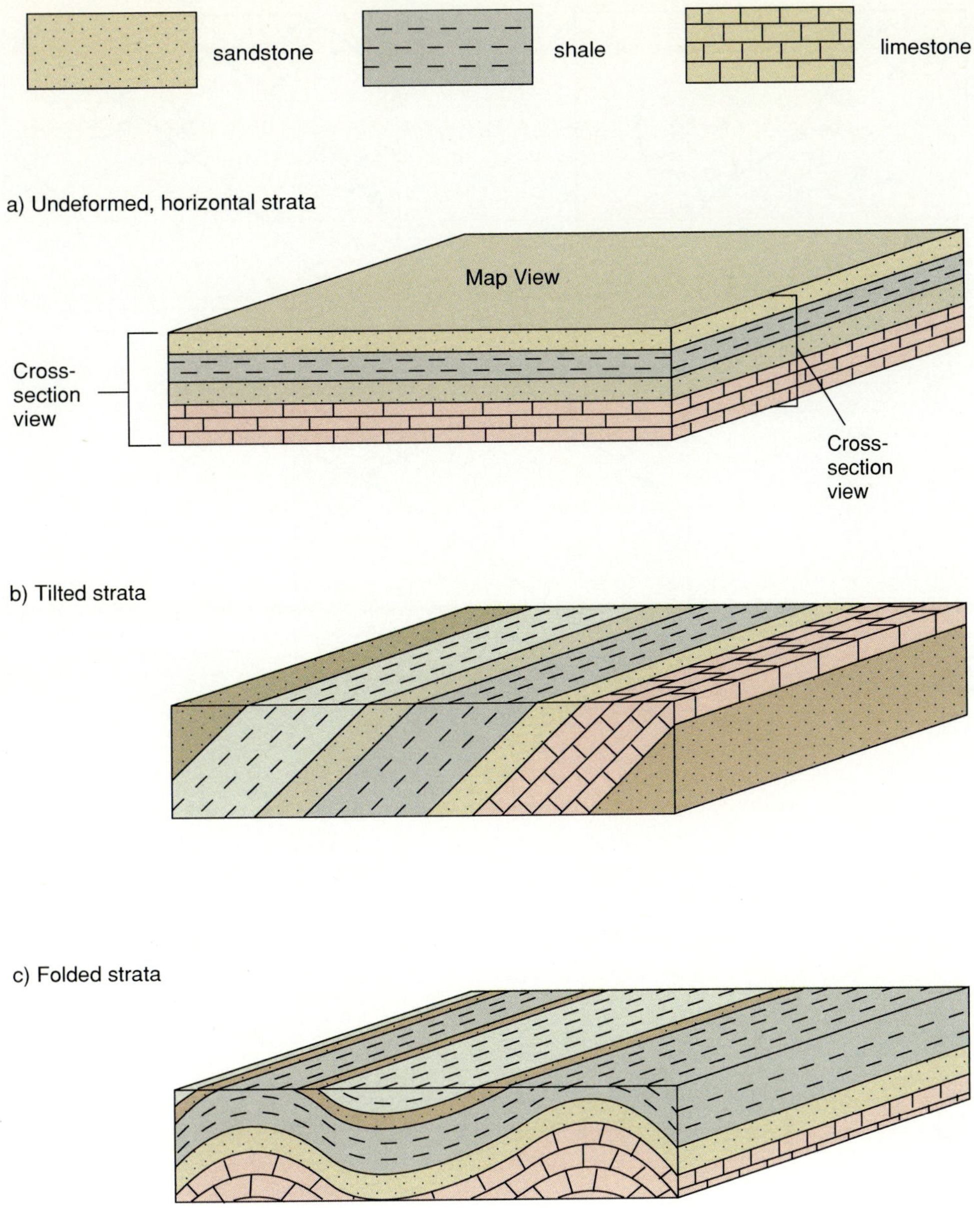

FIGURE 12.1 Block diagrams showing deformation.

---

f) In which areas would you expect the erodability of rock to significantly influence where streams will develop? Explain.

Figure 12.2 shows five common stream drainage patterns. The arrows indicate the direction of stream flow. Match each pattern with one of the following structural situations, and explain your reasoning.

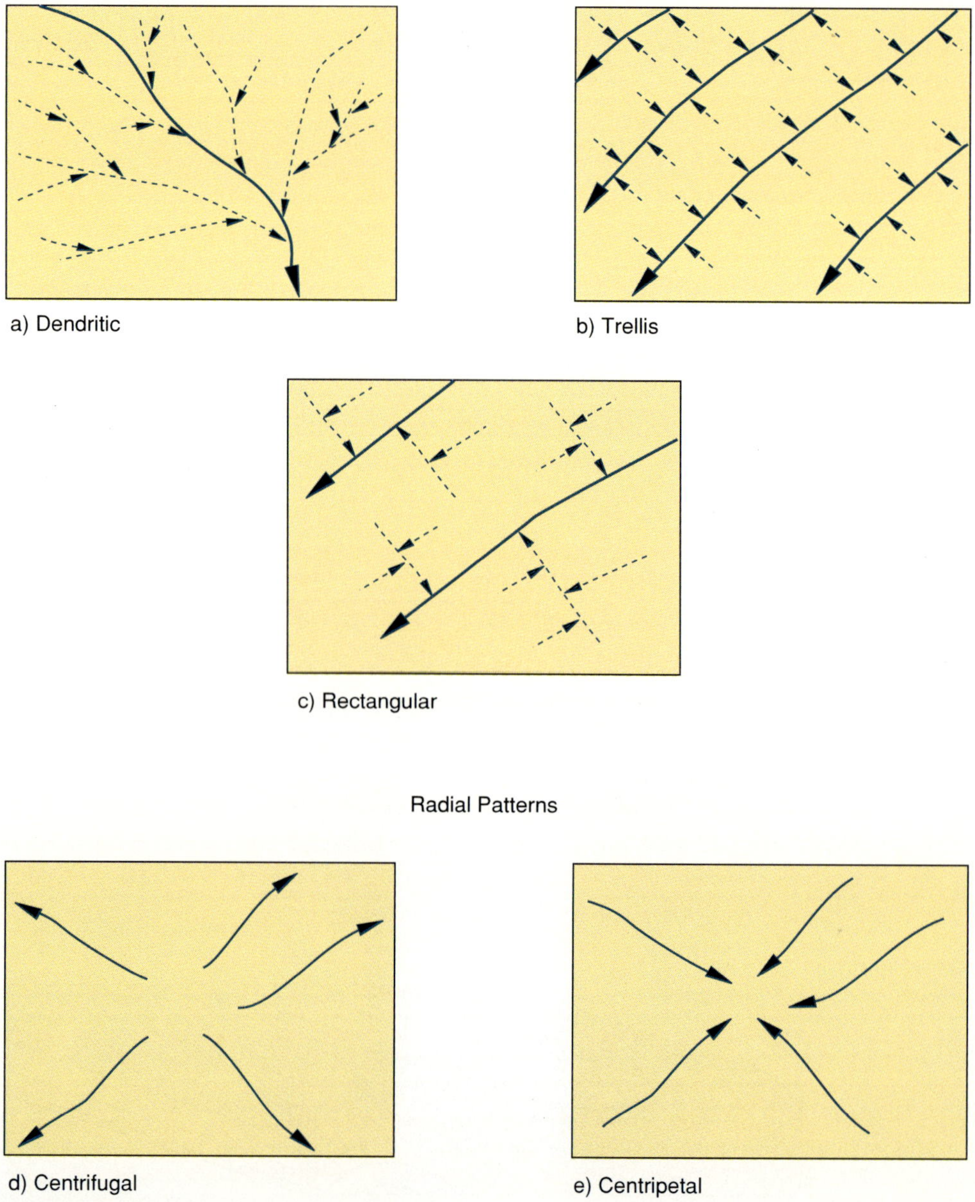

FIGURE 12.2 Drainage patterns. Heavy lines represent major streams; broken lines represent tributaries.

---

g) Describe how the pattern results from the particular state of deformation or other controlling factor. [HINT: In which areas are the streams most randomly oriented? In which are they arranged in the most orderly fashion?]

Undeformed, horizontal strata: ____________________ pattern.

Tilted or folded strata: ____________________ pattern.

A large basin or meteorite impact crater: ____________________ pattern.

A large domal structure or volcano: ______________ pattern.

An area characterized by several faults arranged at nearly right angles to one another: ______________ pattern.

EXERCISE 12.2 MAKING A GEOLOGIC MAP FROM TOPOGRAPHIC CLUES

Figure 12.3a and b show the topography of areas that have experienced quite different histories of deformation. You should be able to recognize the differences and propose a hypothesis explaining what has happened in each area.

a) Place a piece of tracing paper over the areas shown in figure 12.3a and b. Trace the streams with a colored pencil, and describe the drainage patterns present in each area.

b) What can you say about the deformation in the two areas?

With a general idea of the type of deformation in each area, it should be possible to construct a geologic map for each. A geologic map shows where different rocks are found. In this case, the differences are revealed by varying resistance to erosion. Keep the following two points in mind as you respond to the instructions below: (1) resistant rocks stand up as ridges, and non-resistant rocks occupy lowlands; (2) abrupt slope changes also indicate differences in erodability.

c) Using a fine-tipped black pen, trace the contacts between different rock types on the two maps.

d) Color the area underlain by the most resistant rock in figure 12.3a. Use two different colors to indicate the less resistant rocks on either side of that rock unit.

e) Select any one rock unit in figure 12.3b and color it in. Use different colors for the adjacent rock units.

You have just constructed two geologic maps, representing your own hypotheses about the types of material that underlie the two different areas. With a little more information, you will be able to interpret the structures in more detail.

W A R N I N G

Do not overgeneralize! Homogeneously erodable rocks exposed at the surface may not be a single layer of sedimentary rocks. They may be part of a batholith or metamorphic rock complex, or they could be several different rock types, all of which happen to erode at the same rate under the local climatic conditions.

FIGURE 12.3 a) Part of the Three Fords Canyon, Utah Quadrangle. Scale: 1:24,000. Contour interval: 40′. b) Part of the Tremont, Pennsylvania Quadrangle. Scale: 1:24,000. Contour interval: 20′.

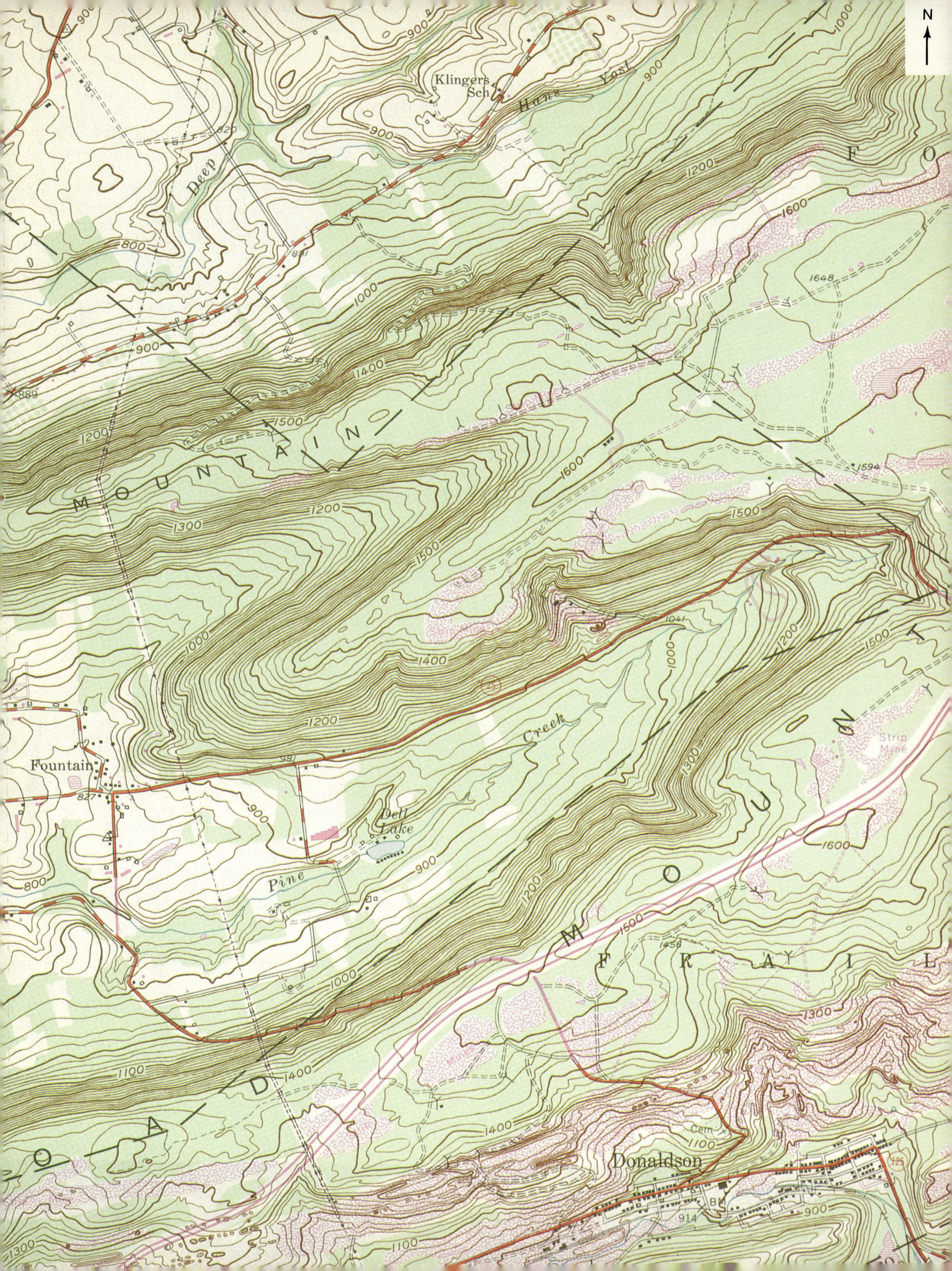
N
Klingers Sch
Deep
MOUNTAIN
Fountain
Dell Lake
Pine
Creek
Strip Mine
MOUNTAIN
BROAD
Donaldson
Cem
BM
914

Folds and Faults

A brief look at figure 12.1b and c suggests that tilted and folded strata yield similar distributions of rocks at the surface: parallel bands of different rock types. A closer look reveals the differences. Folds, tilted beds, and faults are characterized by distinctive outcrop patterns that permit rapid recognition of the type of structure present. We will look briefly at the origin of these outcrop patterns and then try to analyze a few selected geologic maps.

Folds A fold formed from a single sedimentary bed consists of two sides (called **limbs**) that meet at a line called the **fold axis** (Figure 12.4). When several layers are folded, their axes all lie in a single plane, called the **axial plane** of the fold. The precise three-dimensional orientation of the layers is described by their **strike** and **dip.** The strike of a bed is the direction of a horizontal line drawn on the bedding plane, while the dip is the amount and direction that the bed has been tilted downward from its original horizontal position.

Figure 12.4 shows what happens to a sequence of sedimentary rocks when folding takes place. Anticlines (upward-arched beds) and synclines (downward-bent beds)

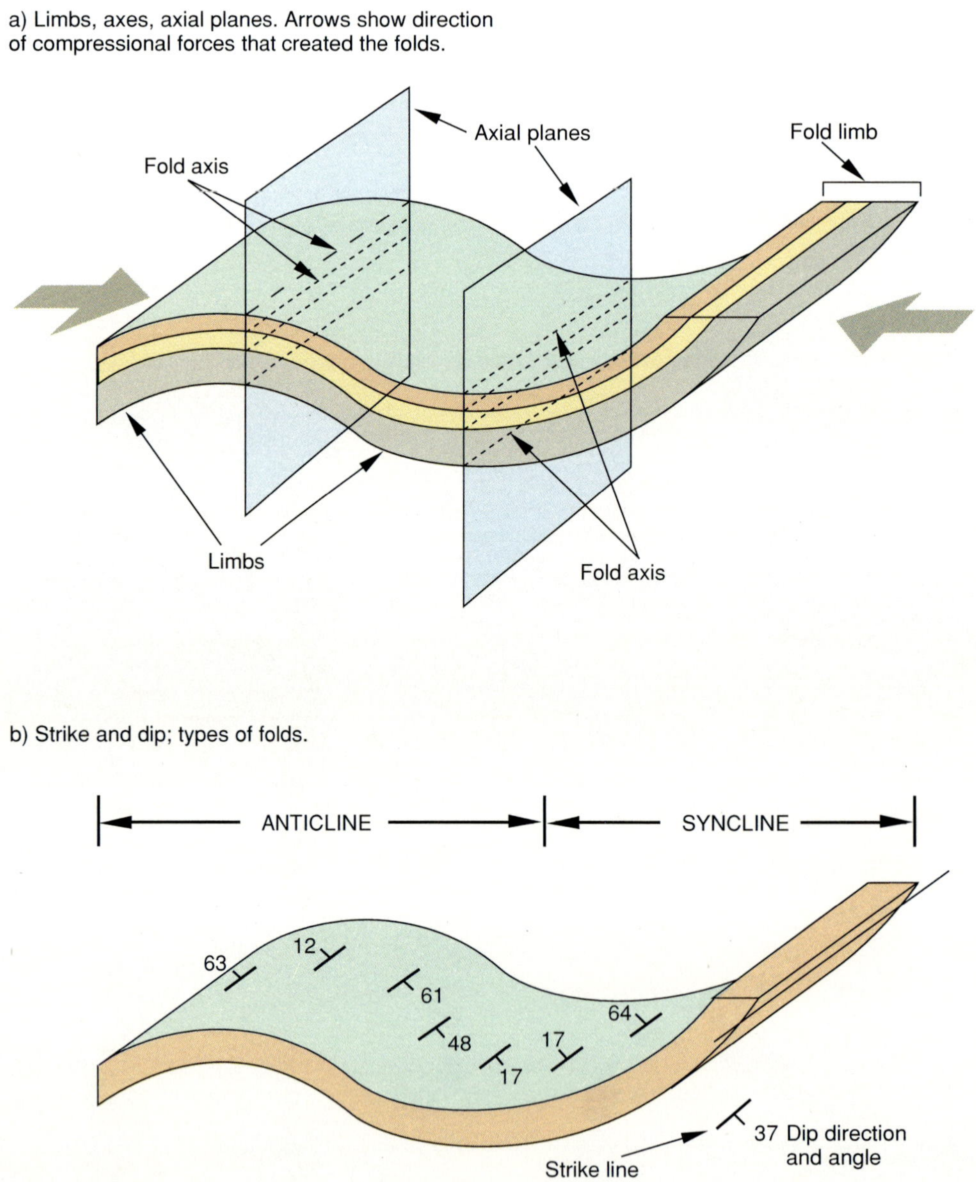

FIGURE 12.4 Fold anatomy and nomenclature.

form axial planes of which are at nearly right angles to the compressional forces responsible for the folding. In the simple structures shown, the limbs of the anticline dip **away from** the axial region of the fold; those of the syncline dip **toward** the axis. Note that the two folds shown in figure 12.4 share a limb. Figure 12.4b shows the strike and dip of one of the layers as it changes orientation in the fold.

Figure 12.5 shows the origin of the outcrop pattern of such folds. For simplicity, the oldest rock is numbered 1, and progressively younger ones, 2 and 3. Compare figure 12.5c with figure 12.1b, the outcrop pattern for tilted strata.

EXERCISE 12.3 ANALYZING GEOLOGIC MAPS

a) What is the difference? (HINT: start at the right side of figure 12.1b, and number the rocks consecutively as you move toward the left.)

b) Why is the outcrop pattern of folded rocks arranged symmetrically on opposite sides of the axis?

Compare figure 12.5c with the geologic map you made on figure 12.3b.

c) How are the elongate bands of rock different on the two maps?

Both areas have been folded, but the zig-zag pattern in figure 12.3b is different from that in figure 12.5c, because the folds present in this area of Pennsylvania have axes that are not horizontal as are those in figures 12.1c and 12.5c. Such folds are called **plunging folds** (Figure 12.5d). Folds with horizontal axes are said to be **non-plunging.**

Faults In contrast to the regular, symmetrical repetition of beds characteristic of the outcrop pattern of folds, faults produce truncation and offset of beds. Where a fault cuts across contacts between layers at a high angle, as viewed on a geologic map, the offsets are easy to spot (Figure 12.6a). Where a fault is parallel to contacts on the surface, the fault may not be as obvious. In such cases, the **asymmetric repetition** or **omission** of units shown in figure 12.6c and d alerts us to the presence of a fault. The offset of rock layers is due to a two-step process: first the uplift of rocks on one side of the fault relative to those on the other, then erosion of the uplifted side more than the downdropped side (because of higher stream gradients). As erosion occurs, the position of the contacts between rock units shifts in the direction of dip, as shown in figure 12.6.

Figure 12.7 shows the four most common types of faults. **Strike-slip faults** are those that have moved horizontally, parallel to the strike of the fault plane, generally as the result of shear stresses. **Dip-slip faults** have undergone some vertical offset, with fault blocks moving up or down the dip of the fault plane. **Normal faults** are dip-slip faults caused by tension. In normal faults, the block resting on the fault plane (the **headwall** or **hanging-wall** block) moves downward relative to the block below the fault plane (the **footwall block**). **Reverse faults** are dip-slip faults caused by compression; their hanging-wall blocks have moved upward relative to their footwall blocks. **Thrust faults** are a type of reverse fault that dips very gently, enabling old rocks to be brought from depth to lie on top of younger rocks in a reversal of the Principle of Superposition.

These are the simplest kinds of faults. Many faults display a combination of dip-slip and strike-slip motion, and others have curved rather than flat surfaces. Yet others have a rotational type of motion rather than the straight-line movement shown. For our purposes, however, we shall deal only with the simplest types.

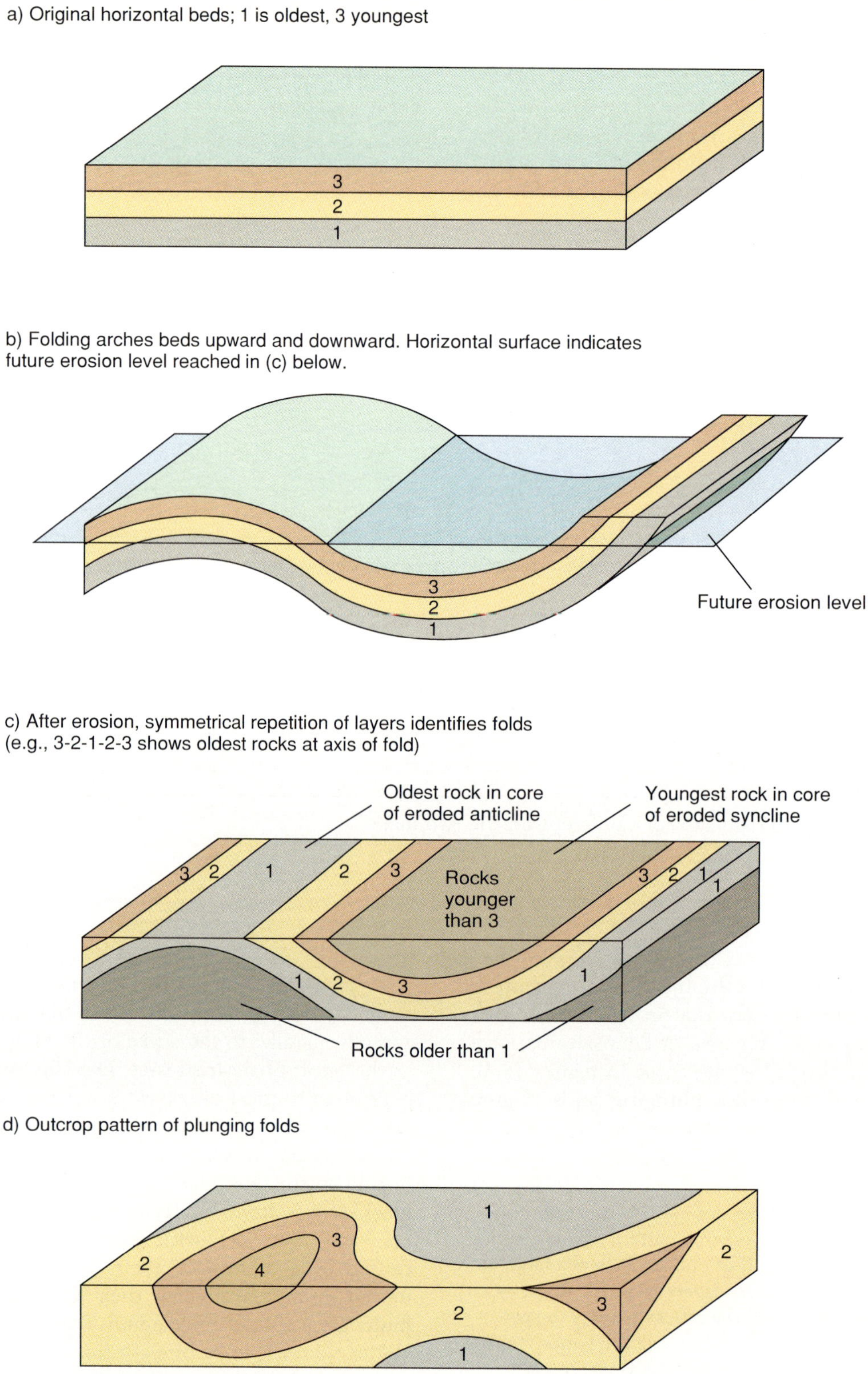

FIGURE 12.5 Recognizing anticlines and synclines.

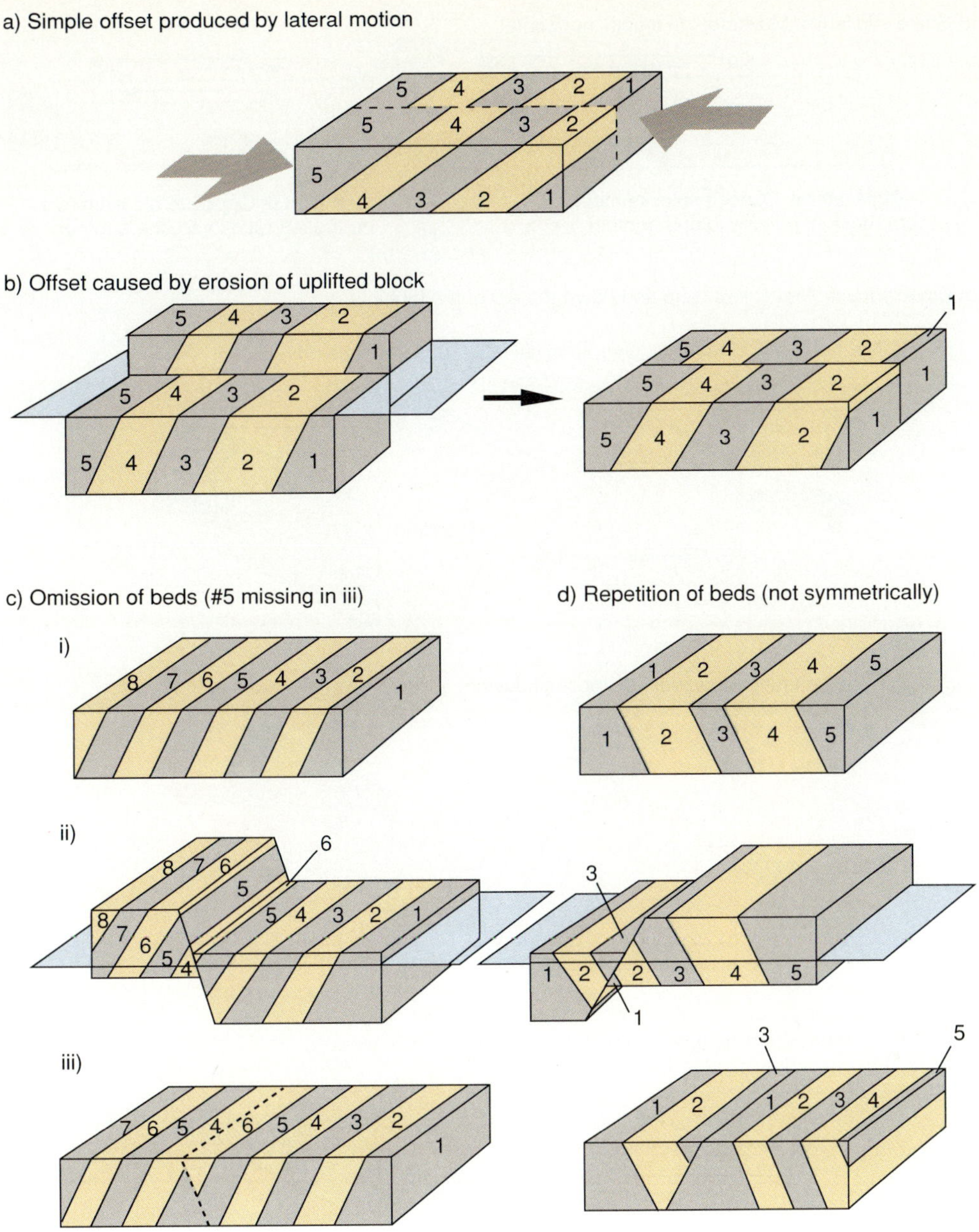

FIGURE 12.6 Outcrop patterns reveal faults.

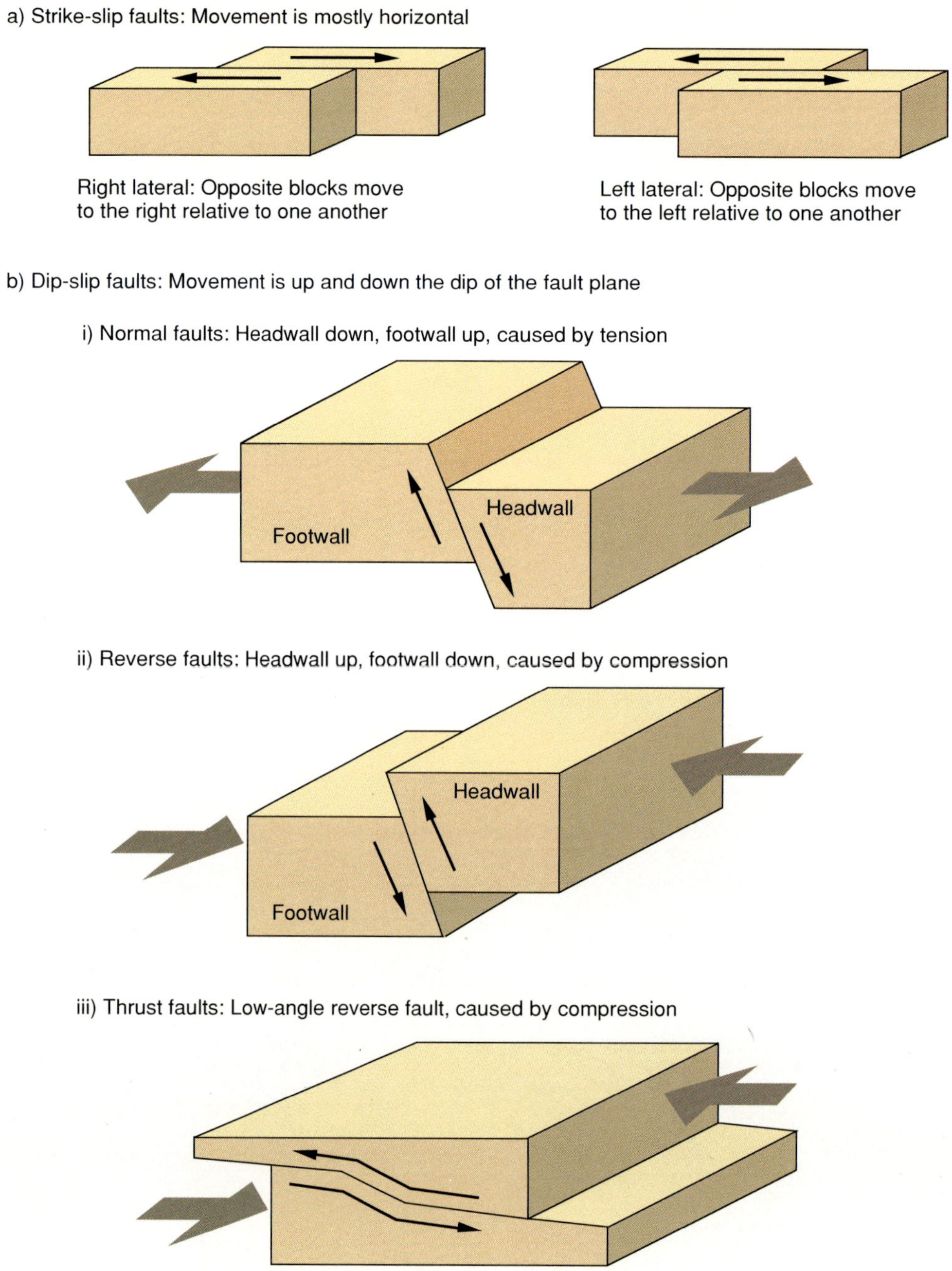

FIGURE 12.7 Classification of faults.

EXERCISE 12.4 COMPLETING BLOCK DIAGRAMS

Figure 12.8 contains several block diagrams in which the map view has been completed by the geologists working in the areas, but the interpretive cross-section views are incomplete. Use the strike and dip symbols to determine the correct dip direction and angle for the layers (as shown in figure 12.8a) and complete each of the diagrams.

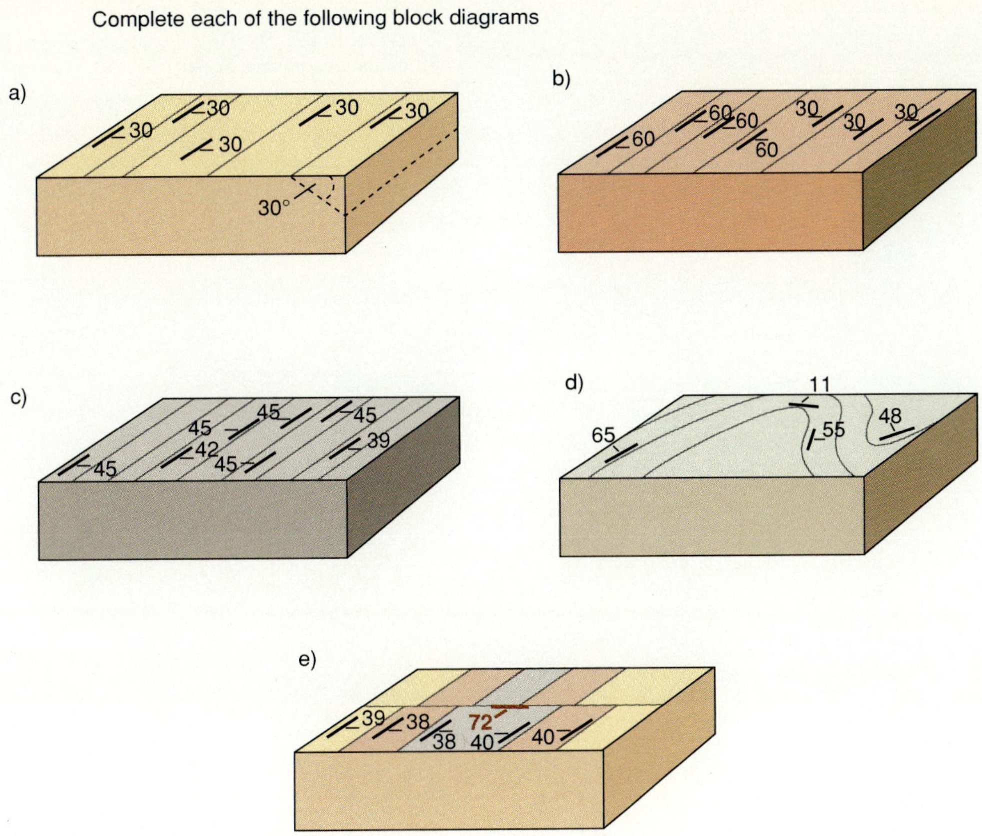

FIGURE 12.8 Block diagram exercises.

Where faults are present, indicate the relative motion of the fault blocks with arrows, and state the type of stress responsible (tension, compression, shear) if possible.

Now let's try it with the real thing! Examine figure 12.9, part of the geologic map of the Devil's Fence Quadrangle, Montana.

---

a) Describe the outcrop pattern present. What type of deformation is indicated? What kind of structure(s) is present?

---

b) Construct a geologic cross section from A to A′. First, draw the **topographic** profile, and then fill in the interpreted geology based on the strike and dip symbols on the map.

Do the same with the areas shown in figures 12.10 and 12.11.

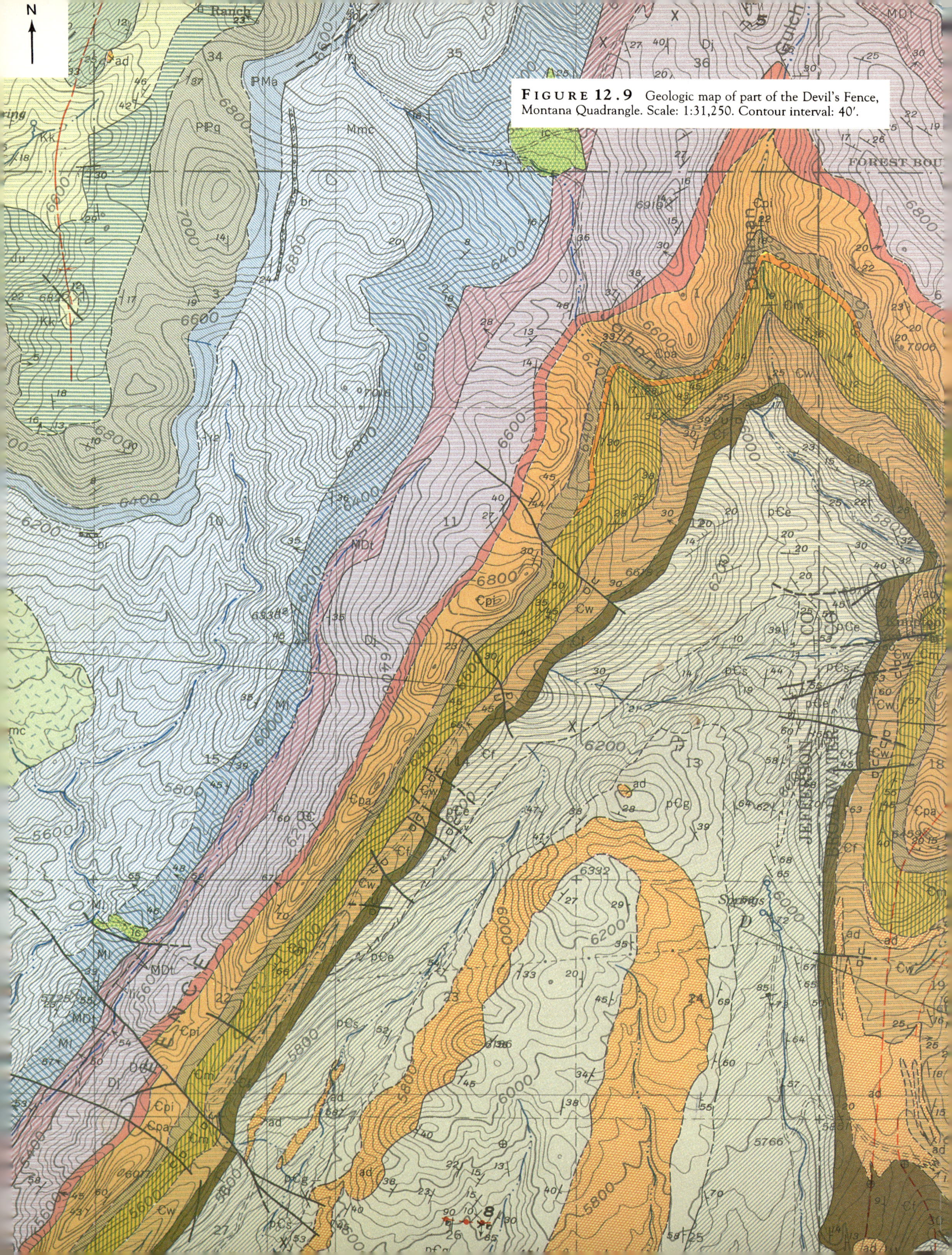

FIGURE 12.9 Geologic map of part of the Devil's Fence, Montana Quadrangle. Scale: 1:31,250. Contour interval: 40′.

EXPLANATION

QUATERNARY — Recent

Qal
Alluvium

TERTIARY — Pliocene(?)

Ttg
Fan gravel and pediment gravel
Unconsolidated or weakly consolidated fan deposits and gravel veneer on pediments

Miocene(?)

Ttg
Tuff and gravel
Light-colored rhyolitic tuff with a few interbeds of stream gravel

UNCONFORMITY

CRETACEOUS — Lower and Upper Cretaceous

Kcm, Kcl
Colorado formation
Upper black shale unit not mapped. Middle siliceous mudstone and sandstone unit, Kcm, is intertongued marine and nonmarine sandstone, mudstone, and siliceous mudstone, in part of volcanic origin. Lower black shale unit, Kcl, is a basal tan-weathering quartz sandstone, drab and olive-gray blocky siltstone, very dark gray to black shale, and an upper dark-gray carbonaceous, limonitic speckled to mottled sandstone

EROSIONAL UNCONFORMITY

Lower Cretaceous

Kk
Kootenai formation
Comprises three units. Upper unit is 10 to 25 ft of gastropod-bearing limestone overlain by a few feet to 80 ft of drab mudstone; middle unit is red and green mudstone and shale with concretions and lentils of limestone; lower unit is crossbedded "pepper-and-salt" sandstone and interbedded shale and mudstone

JURASSIC — Upper Jurassic

Ju
Morrison and Swift formations
Morrison formation: varicolored nonmarine shale, mudstone, and siltstone, with thin beds of limestone and sandstone, and near the top a unit of black shale. Upper hundred feet locally contains thick, lenticular "pepper-and-salt" sandstone and grades into overlying Kootenai formation
Swift formation: grayish-brown punky calcareous marine sandstone, 20 to 35 ft thick, with a basal chert-pebble conglomerate

EROSIONAL UNCONFORMITY

PERMIAN

Pp, PPq
Phosphoria formation
Brown and gray chert and sandstone, in part phosphatic; may locally contain one or two thin beds of phosphate rock, Pp. In places mapped with the Quadrant formation, PPq

CARBONIFEROUS — PENNSYLVANIAN

Pq
Quadrant formation
Light-colored quartzitic sandstone and interbedded light-gray sugary-textured sandy dolomite

PMa
Amsden formation
Red to grayish-red mudstone, shale, and subordinate amounts of carbonate rock with interbeds of gray, brown, or yellow argillaceous sandstone in upper and lower parts; middle part of medium- to dark-gray thick-bedded dolomite

EROSIONAL UNCONFORMITY(?)

MISSISSIPPIAN — Madison group

br, Mmc
Mission Canyon limestone
Medium-gray to light-gray medium-grained thickly and indistinctly bedded limestone, with a few thin siliceous layers in lower 200 ft and sparse gray chert nodules and lentils in upper half. A breccia unit, br, about 200 ft below top of formation has been mapped locally

Ml
Lodgepole limestone
Upper part of medium-gray fine- to medium-grained limestone in distinct beds as much as 3 ft thick alternating with zones of much thinner beds containing rare mudstone partings; lower part of medium-gray limestone in beds 1 in. to 1 ft thick with partings and interbeds of yellow to red calcareous mudstone; grades into Mission Canyon limestone through a 150- to 200-ft zone

DEVONIAN AND MISSISSIPPIAN — Upper Devonian and Mississippian

MDt, tr
Three Forks shale
Predominantly greenish-gray and brown shale with subordinate amounts of interbedded sandstone and limestone. **Dolomitic** *siltstone at top. Locally a 10- to 25-ft fossiliferous limestone unit, tr, has been mapped*

DEVONIAN — Upper Devonian

Dj
Jefferson dolomite
Dark-gray granular-weathering fetid well-bedded dolomite with subordinate amounts of dark-gray limestone and light-gray dolomite

Upper Cambrian and Upper Devonian

EROSIONAL — — — — DЄ — — — *UNCONFORMITY*

Maywood and Red Lion formations undifferentiated
Varicolored, generally in shades of red and yellowish-brown, argillaceous, dolomitic, and calcareous rocks; poorly exposed

CAMBRIAN — Upper Cambrian

Єpi
Pilgrim dolomite
Comprises three units. Upper unit is light-gray thick-bedded dolomite commonly mottled medium-gray near base. Middle unit is light- to medium-gray crystalline limestone irregularly ribboned with yellowish-gray silty dolomite. Lower unit is mottled light- and dark-gray dolomite with sparse intraformational conglomerate; locally, basal 8 to 10 ft is bluish-gray limestone

Middle Cambrian

Єpa
Park shale
Olive-gray, gray, and light-brown shale with minor amounts of argillaceous limestone, siltstone, and sandstone

Єm
Meagher limestone
Comprises three units. Upper and lower units are medium-gray limestone irregularly ribboned or mottled with yellowish-orange, yellowish-brown, and yellowish-gray dolomite. Middle unit is thickly and indistinctly bedded medium-gray limestone, commonly with oolitic beds

Єw
Wolsey shale
Upper half is interbedded gray argillaceous limestone and greenish- and yellowish-gray calcareous mudstone and shale. Lower half is greenish-gray and drab shale with some interbeds of sandstone and limestone; many beds are micaceous, some are glauconitic

Єf
Flathead quartzite
White to pale shades of gray, pink, brown, and purple medium- to thick-bedded homogenous even-grained quartz sandstone; most beds are cemented to vitreous quartzite; thin, discontinuous sparse pebble zones in lower part; crossbedding common

UNCONFORMITY

PRECAMBRIAN — Upper Precambrian — Belt series

pЄe
Empire shale
Gray, greenish-gray, and brown, siliceous mudstone or argillite with interbeds of quartzite sandstone and shale. Intertongues with Spokane shale

pЄs
Spokane shale
Grayish-red mudstone, shale, and sandstone, with a few thin beds of limestone near base

pЄg
Greyson shale
Gray and brown mudstone or shale alternating with sandstone or quartzite. Base not exposed. Grades into Spokane shale

YOUNGER INTRUSIVE ROCKS

TERTIARY(?) — Lower Tertiary(?)

gd
Granodiorite and quartz diorite
In Sagebrush Park stock

ic
Composite or hybrid intrusives
Small plutons containing diverse and unusual rocks including olivine-rich and quartz-rich types

OLDER INTRUSIVE ROCKS

CRETACEOUS — Upper Cretaceous

ba
Basalt and related rocks
Dark-gray to greenish-black fine- to medium-grained rocks, mainly as sills

ad
Andesite porphyry, diorite porphyry, and related rocks
Greenish-gray to dark-gray porphyritic rocks with phenocrysts of plagioclase and hornblende or augite; mainly as sills

hl
Hornblende lamprophyre
Very fine grained gray rock with conspicuous hornblende phenocrysts

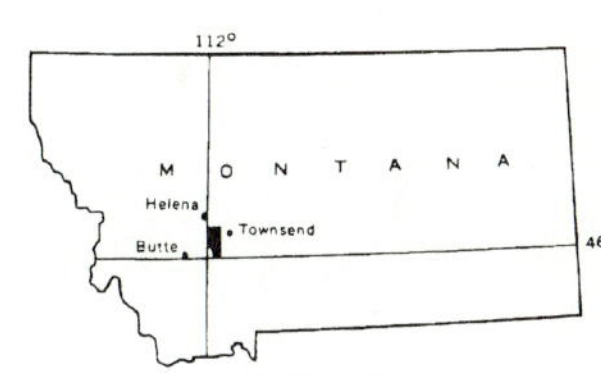

INDEX MAP

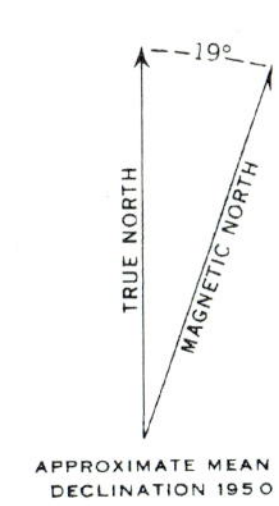

APPROXIMATE MEAN DECLINATION 1950

67
Contact, showing dip
Dashed where approximately located; length of dashes indicates relative degree of accuracy; long dash most accurate

90
Vertical contact

Overturned contact

U 39 ? D
Fault, showing dip
Dashed where approximately located; length of dashes indicates relative degree of accuracy; long dash most accurate. Dotted where concealed, queried where doubtful. U, *upthrown side;* D, *downthrown side*

F
Preintrusion fault
Forming boundary along margin of intrusive body; dashed where approximately located

Anticline
Showing crestline and plunge of axis. Dashed where approximately located; length of dashes indicates relative degree of accuracy; long dash most accurate. Dotted where concealed; queried where doubtful

Syncline
Showing position of trough and plunge of axis. Dashed where approximately located; length of dashes indicates degree of accuracy; long dash most accurate. Dotted where concealed; queried where doubtful

90
Vertical vein, approximately located

85
Vein, approximately located, showing dip

Vertical shaft

Adit

Prospect pit

Mine

Overturned anticline

13
Strike and dip of beds

90
Strike of vertical beds

Horizontal beds

50
Strike and dip of cleavage

17
Strike and dip of joints

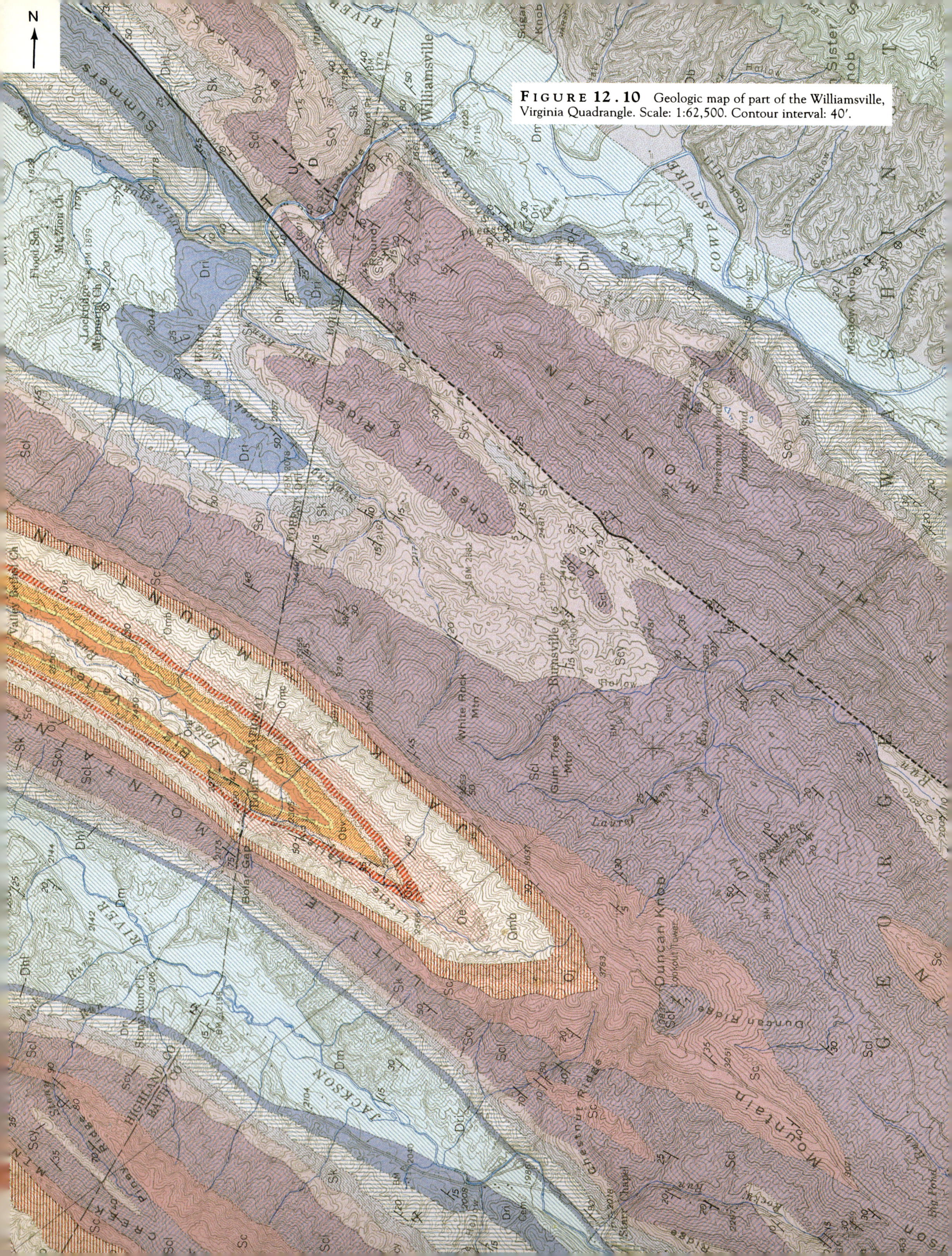

FIGURE 12.10 Geologic map of part of the Williamsville, Virginia Quadrangle. Scale: 1:62,500. Contour interval: 40′.

EXPLANATION

DEVONIAN

Upper Devonian

Dch — Chemung formation

Middle Devonian

Db — Brallier shale

Dm — Millboro shale (*including Onondaga fm. in southeast*)

Lower Devonian

Dri — Ridgeley sandstone (*including Onondaga fm. in northwest*)

Dhl — Helderberg group (*Coeymans ls.; New Scotland ls.; Licking Creek ls.*)

SILURIAN

Upper Silurian

Keyser limestone

Scy — Cayuga group (*McKensie ls.; Wills Creek fm.; Tonoloway ls.*)

Middle Silurian

Scl — Clinton formation

Lower Silurian

Sc — Clinch sandstone

ORDOVICIAN

Upper Ordovician

Oj — Juniata formation

Middle Ordovician

Omb — Martinsburg formation

Oe — Edinburg formation (*Lantz Mill facies*)

Om — Moccasin formation

Omc — McGlone formation

Obv — Big Valley formation

Ol — Lincolnshire limestone

Olu — Lurich formation (*Basal clastics; Five Oaks mem.*)

Lower Ord.

Ob — Beekmantown dolomite

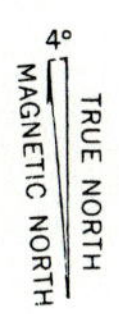

APPROXIMATE MEAN DECLINATION, 1946

FAULTS
NORMAL AND REVERSE

U/D (solid line) — exposed

U/D (dashed line) — approximate

U - *upthrown side*

D - *downthrown side*

CONTACTS

exposed

approximate

covered

ATTITUDE OF ROCKS

60 — Strike and dip of beds

60 — Strike and dip of overturned beds

Strike of vertical beds

⊕ Horizontal beds

THRUST

exposed

approximate

T overthrust side

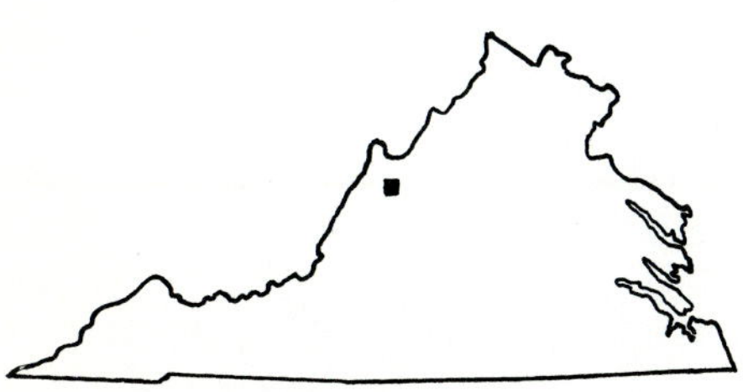

QUADRANGLE LOCATION

SCALE 1:62500

1 ½ 0 1 2 3 4 MILES

3000 0 3000 6000 9000 12000 15000 18000 21000 FEET

CONTOUR INTERVAL 40 FEET
DATUM IS MEAN SEA LEVEL

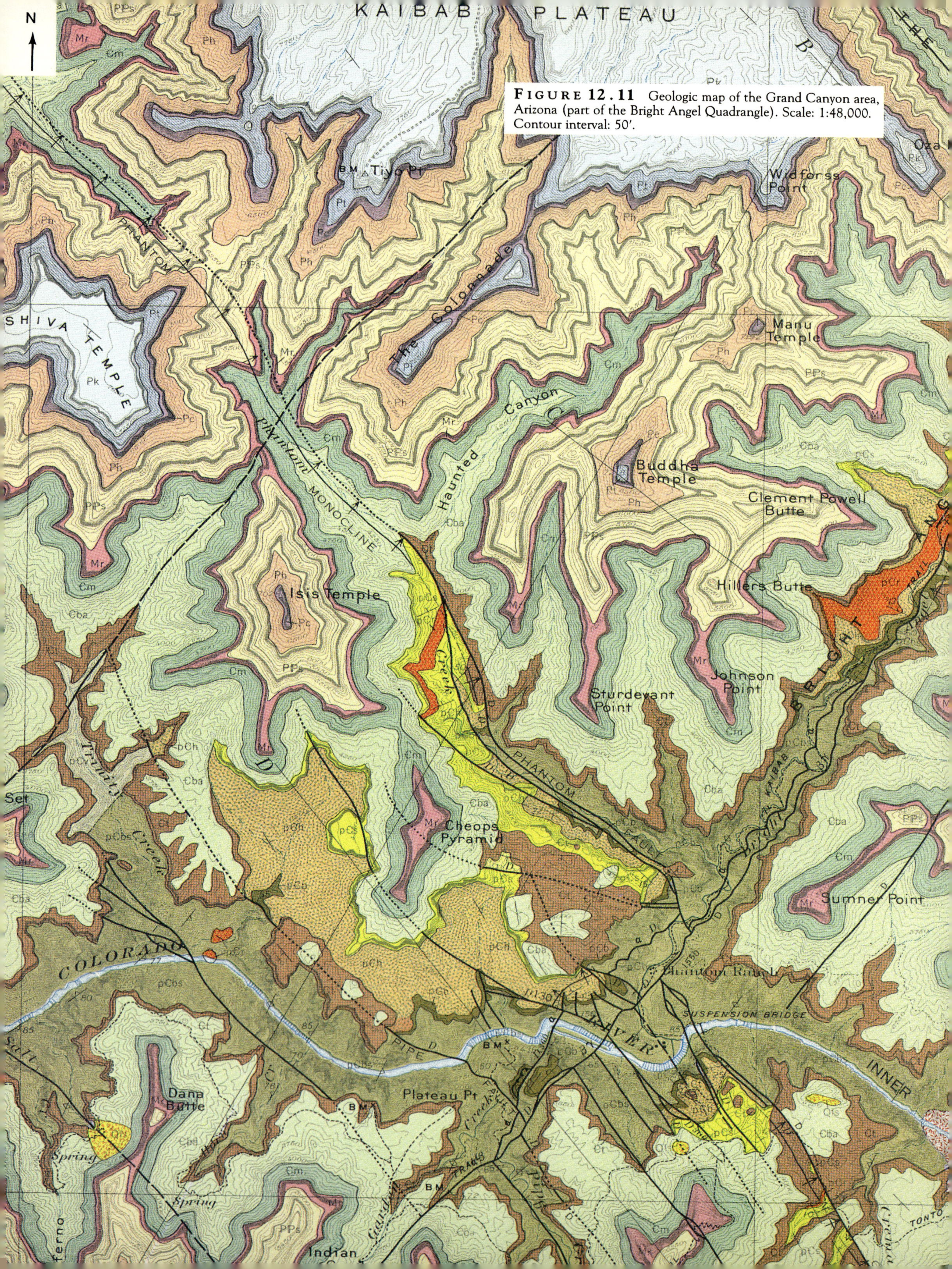

FIGURE 12.11 Geologic map of the Grand Canyon area, Arizona (part of the Bright Angel Quadrangle). Scale: 1:48,000. Contour interval: 50′.

EXPLANATION

CENOZOIC

Quaternary

- Qls — Landslide

Laramide Revolution

MESOZOIC

Triassic

- T̄Rm — Moenkopi Formation

PALEOZOIC

Permian

- Pk — Kaibab Formation
- Pt — Toroweap Formation
- Pc — Coconino Sandstone
- Ph — Hermit Shale

Permian and Pennsylvanian

- PIPs — Supai Formation

Mississippian

- Mr — Redwall Limestone

Cambrian

- Ꞓm — Muav Formation
  Includes unclassified Middle Cambrian strata at top. Devonian Temple Butte Limestone occurs here and there filling channels eroded in upper part of the formation.
- Ꞓba — Bright Angel Formation
- Ꞓt — Tapeats Sandstone

Grand Canyon Revolution and Grand Canyon Peneplain.

ALGONKIAN

Upper Precambrian

- pꞒr — Rama Formation
  Diabase plugs, dikes, and sills intrusive into older formations.
- pꞒd — Dox Formation
- pꞒs — Shinumo Sandstone
- pꞒh — Hakatai Shale
- pꞒb — Bass Limestone

Arizonan Revolution and Arizonan Plain.

ARCHEAN

Lower Precambrian

- pꞒz — Zoroaster Granite
  Stocks and pegmatite dikes intrusive into older formations.
- pꞒbs — Brahma Schist
  Principally metamorphosed basic rocks presumably of volcanic origin.
- pꞒv — Vishnu Schist
  Meta-sediments with minor meta-volcanics.

Note: The terms Archean and Algonkian are used as subdivisions of Precambrian time. Their long established use in describing the classic geology of Grand Canyon makes this desirable. The United States Geological Survey and the United States National Park Service do not recognize any formal subdivisions of Precambrian time.

Contact

Gradational contact

Fault

Dashed where approximately located; dotted where inferred beneath Paleozoic strata. Direction and amount of dip indicated.

Ep-Algonkian Fault

D, downthrown side and amount of displacement.

Laramide Fault

D, downthrown side and amount of displacement.

α, first Ep-Algonkian Bright Angel thrust fault.

β, second Ep-Algonkian Bright Angel thrust fault.

γ, Laramide Bright Angel normal (oblique slip) fault.

Fault with strike slip component

Bedding plane overthrust fault

Monocline

Arrows show direction of inclination. Dashed where inferred beneath Paleozoic strata.

Strike and dip of beds

Strike and dip of overturned beds

Strike and dip of foliation

Strike of vertical foliation

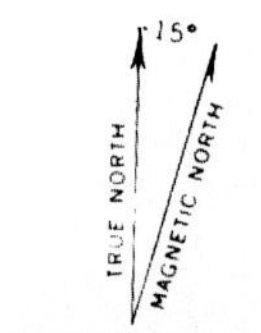

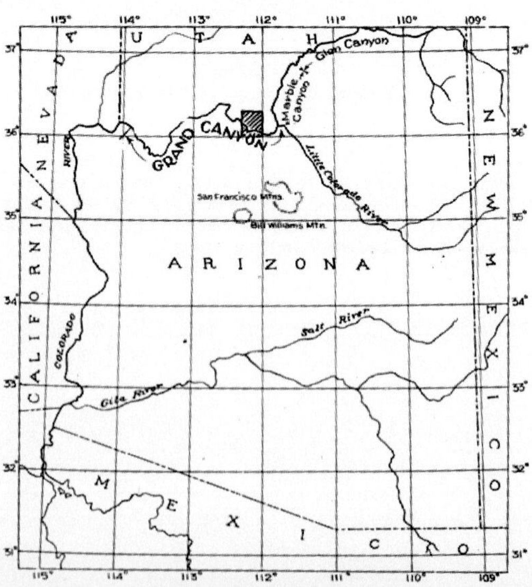

CHAPTER 13

INTERPRETING GEOLOGIC HISTORY

PURPOSE

To practice interpretation of the geologic history of an area from clues provided by topography, lithology, and geologic structures.

MATERIALS NEEDED

1) Colored pencils.
2) A geologic time chart (if you haven't memorized it yet).

INTRODUCTION

The goal of a geologist working in an area is to reconstruct its geologic history. Eventually, by piecing together many local histories, geologists arrive at their ultimate goal: hypotheses concerning the origin and evolution of the entire Earth. The best way for you to learn how to construct such a geologic history is to go out in the field, to a place where you can find bedrock outcrops; then locate them on a map, identify the rocks, see for yourself how the rock units are related to each other, and make your own geologic map and cross sections. The next best method is to complete a series of exercises that leads you through the same thought processes, but without the enjoyment of walking across the actual area.

Previous chapters focused on materials and processes. You learned how to distinguish sandstone from schist, recognize that folding had taken place, and deduce which agents of erosion and deposition had been active in a given area. These skills helped you to build a three-dimensional picture of the Earth. In this chapter, we add the fourth dimension—**time.** You will draw on the skills that you have gained from the preceding chapters to unravel the sequence of events that has occurred in each of the areas shown in the following exercises. In doing so, you will be determining the **relative ages** of each rock unit and geologic event. In many instances, it may not be apparent precisely when the events occurred in geologic time, but you will be able to ascertain their correct order.

EXERCISE 13.1 CRITERIA FOR RELATIVE AGE DATING

Several principles permit geologists to determine the relative ages of materials and processes. Most of these "principles" are little more than common sense. Some deal specifically with sedimentary rocks, and others with any kind of rock. We will start with a basic assumption about sedimentary rocks, and proceed from there. The **Principle of Original Horizontality** states that most sedimentary rocks are deposited as horizontal beds.

---

a) Re-examine figure 10.2, a view of the Grand Canyon. Assuming original horizontality, where are the oldest rocks? The youngest? Where should the oldest rocks be in any pile of sedimentary rocks? Explain your reasoning.

The same reasoning was first used more than 300 years ago, and was called the **Principle of Superposition.** It holds for all sedimentary rocks that have not been disturbed by deformation.

Re-examine figure 4.4b, the cross-bedding in Zion National Park. The person in the photo is pointing to a contact that cuts across several beds. The truncation alone demonstrates the relative ages of the materials.

---

b) Which beds must be older: those cut by the truncating layer or the truncating layer itself? Explain.

Right again! This is the **Principle of Cross-Cutting Relations,** which is equally useful for several different kinds of geologic materials and processes (Figure 13.1). An intrusive igneous rock that cuts across other rock units (Figure 13.1a) must be ________ than the units it cuts across. A fault (Figure 13.1b) must be ________ than the rocks it affects. Folds (Figure 13.1c) must be ________________ than all the rocks involved in the folding event.

Examine figure 13.2, a photograph of a granitic intrusive containing fragments of the rock that it intruded.

---

c) Which is older, the granite or the fragments? Explain.

Again, this reasoning is given a formal name: the **Principle of Inclusions.** It works equally well for clastic sedimentary rocks: rock fragments found within a conglomerate must be ________ than the conglomerate itself.

There are several other techniques for relative age dating, although these are not readily visible in the photographs, diagrams, and maps that we will use in the following exercises. The **Principle of Faunal and Floral Succession** states that fossils record a progressive change in the complexity of life forms as geologic time unfolded, with the more complex creatures arising later in time.

Special fossils called **index fossils** are particularly useful because they indicate the precise ages of the strata in which they are found. Index fossils are the remains of plants or animals that had a wide geographic distribution (thus making them easy to find anywhere) but lived for only a short span of geologic time. Each index fossil thus indicates a unique, narrowly constrained age for the rocks in which it is found.

Finally, several sedimentary structures discussed in chapter 4 are useful in relative age dating, because they reveal where the top and bottom of their beds were at the time of deposition. These structures include mudcracks, graded beds, oscillation ripplemarks, and raindrop impressions. If beds have been tilted or folded so that superposition can no longer be applied, or even if they have been completely overturned, we can still recognize which layers are older and which younger. Figure 13.3 shows how these "top-and-bottom structures" can be used.

The value of these top-and-bottom structures is shown in figure 13.4. Examine this block diagram carefully.

a) Intrusive igneous rock, Sierra Nevada mountains, Yosemite National Park, California

b) Fault cutting strata at Kingman, Arizona

c) Fold deforming strata, Southern Utah.

FIGURE 13.1 Cross-cutting relations.

FIGURE 13.2 The Principle of Inclusions. The granite (light colored) has incorporated fragments of the rock that it intrudes. These xenoliths must be older than the granite itself in order to have been included during intrusion (Baring granite, Calais, Maine).

a) Why is it not possible to apply the Principle of Superposition to these sedimentary rocks?

b) Use the sedimentary structures indicated to determine the relative ages of the strata. Number the oldest #1, the next younger #2, and so on. Explain your reasoning.

c) Sketch the appearance in both cross sections of graded bedding in the sandstone layer where it has been repeated by folding.

d) What if the other layer adjacent to the ripplemarked bed was a mudcracked shale? Sketch what you think the mudcracks would look like in the cross-section views.

Figures 13.5 and 13.6 are cross sections of areas with different geologic histories. Use the principles discussed earlier to determine the exact sequence of events in each diagram. Letters are provided to help identify individual features, but they have no significance as far as sequence is concerned; you must arrange the events in the order in which they occurred. For every hypothetical sequence, you must carefully explain your evidence and reasoning.

Evidence may not always be sufficient to make a positive determination of age. In such cases, you will have to make assumptions about the ages of two materials or events, and you may have to present multiple working hypotheses—two or more models for the sequence. Describe any such assumptions, and indicate what additional evidence would enable you to choose between the possible histories.

"Events" include such things as deposition of a sedimentary rock unit, an episode of deformation (tilting, folding, or faulting), intrusion of igneous rocks, and weathering and erosion.

Most of the lines separating rock units in figures 13.5 and 13.6 are contacts between sedimentary beds deposited more or less continuously, and thus **conformably,** upon one another, or are contacts between igneous rocks and the host rocks that they intrude. Some, however, represent a special kind of contact called an **unconformity.** An unconformity is a contact between rock units that represents an interruption in the conformable sequence that one would expect if sedimentation had been continuous. Unconformities indicate a gap in the geologic record provided by the rocks—a time span in which rocks either were not deposited or were subsequently eroded away after their formation.

Unconformities commonly result from major episodes of deformation, following which, uplifted areas are eroded downward (removing some of the rocks that had been deformed), and lower areas are filled in with the eroded debris. The tilted and folded strata are truncated by erosion, and a horizontal blanket of sediment covers the truncated surface. This type of unconformity is called an **angular unconformity** because of the angular discordance between the horizontal rocks above the unconformity and the dipping rocks below it.

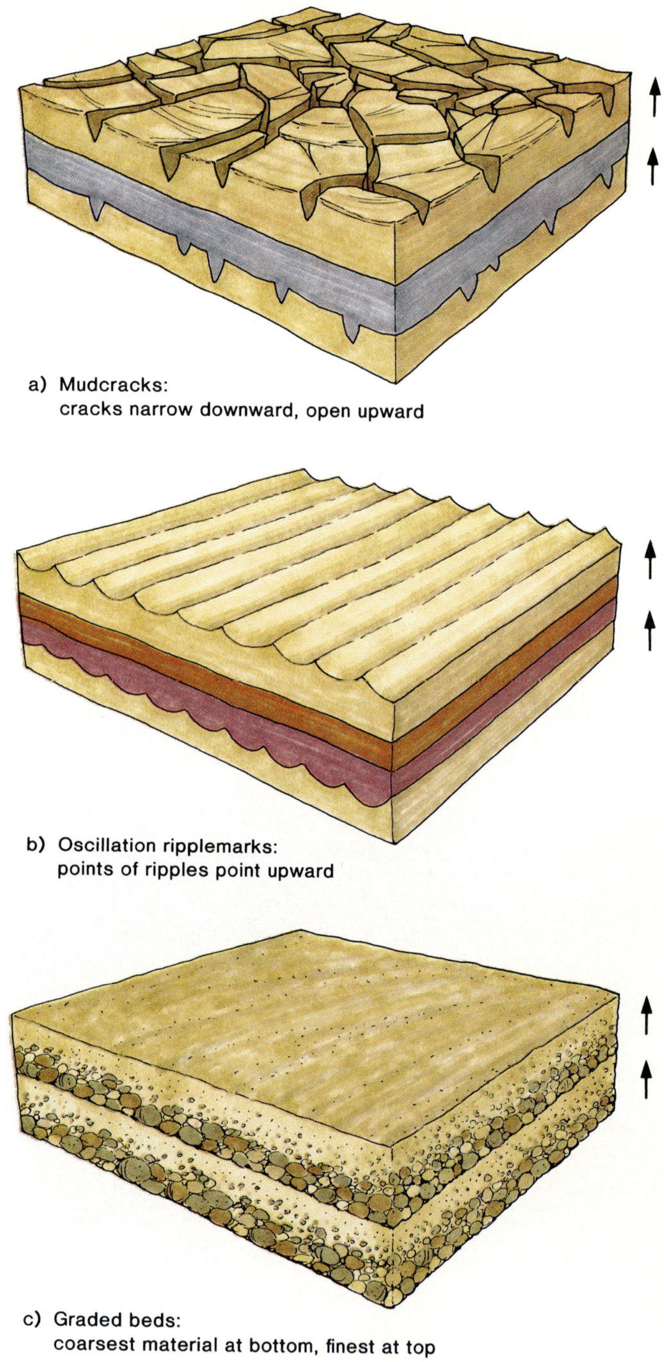

FIGURE 13.3 Top-and-bottom sedimentary structures (arrows point toward top of bed in all cases).

Other types of unconformities may separate rocks that have not undergone intense deformation, and whose beds may thus be parallel to one another. This situation may occur when the base level for streams in an area drops, permitting erosion of previously deposited materials.

Examine figures 13.5 and 13.6. Identify any unconformities that may be present. Which of these unconformities are (is) angular unconformities? Explain how the presence of an unconformity makes it more difficult to unravel the geologic history of an area.

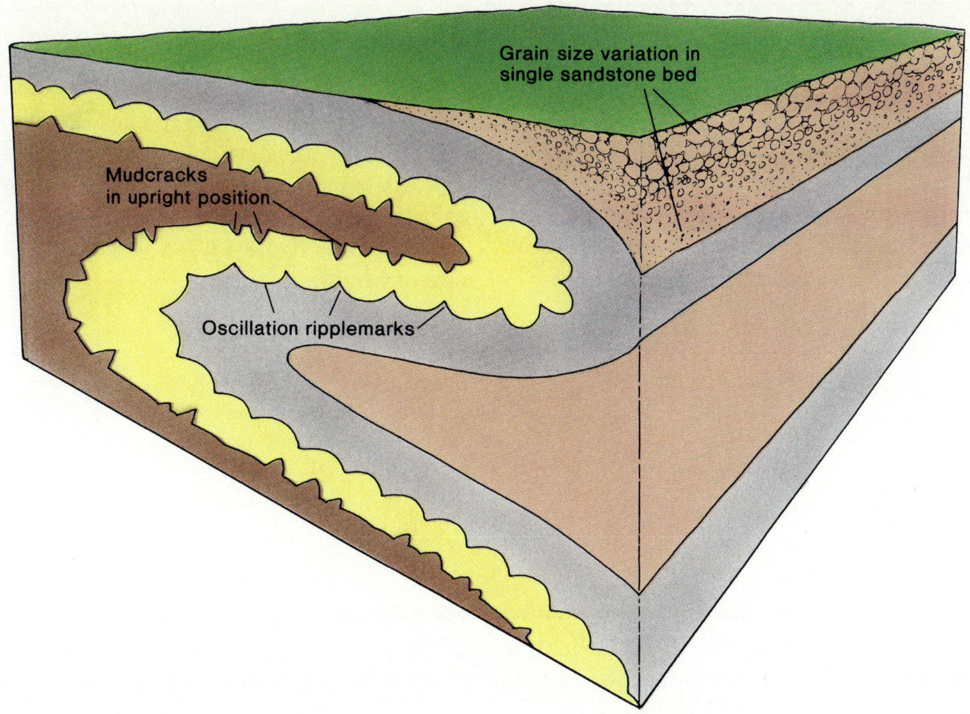

FIGURE 13.4 Use of top-and-bottom structures to interpret relative ages in a deformed sequence of sedimentary rocks.

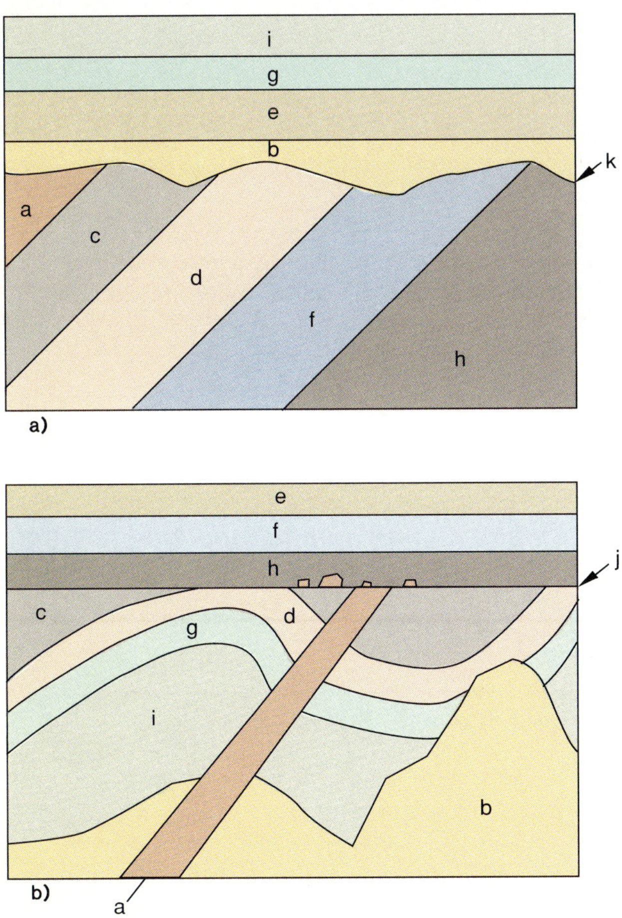

FIGURE 13.5 Beginning exercises in geologic history.

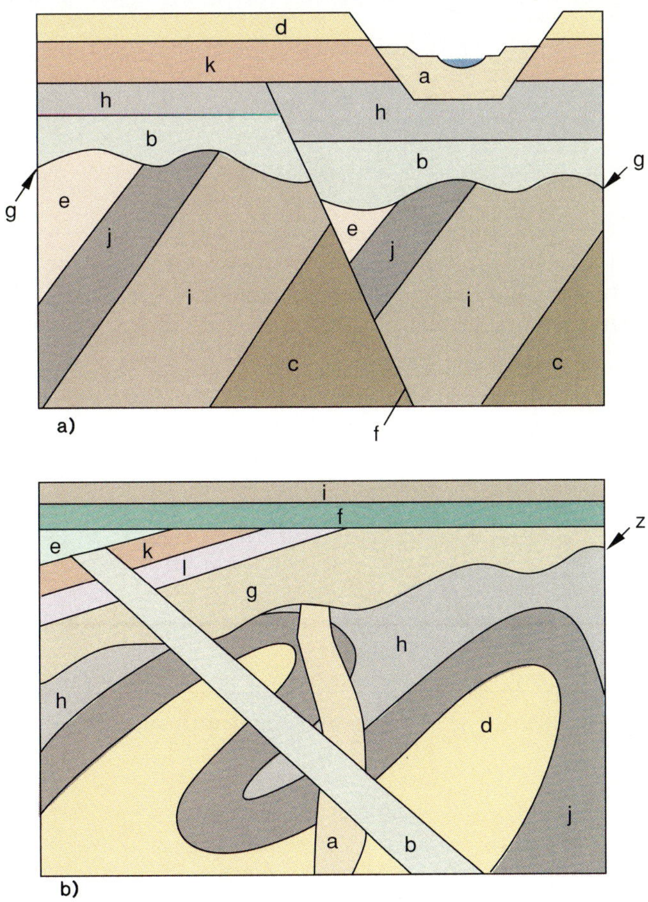

FIGURE 13.6 More advanced exercises in geologic history.

EXERCISE 13.2 INTERPRETING THE HISTORY OF PARTS OF NORTH AMERICA

With all this training, you should now be able to look at a geologic map and come to some conclusions about the history of the region shown. The geologic map of the United States is an excellent place to start, so that you can concentrate on the history of areas near your home and school. The following diagrams introduce you to the types of questions that you can ask and the answers that are possible. If your lab has a geologic map of the entire United States, or of North America, see how each of the areas used below fits into the larger continental context.

Examine figure 13.7, a geologic map of most of the state of Michigan.

a) Describe the outcrop pattern present.

b) The topographic basins that today form Lakes Michigan and Huron and Saginaw Bay were carved by Pleistocene continental glaciation, and the water today interrupts the otherwise continuous outcrop patterns of the rock units. Use a pencil to indicate with dotted lines where you think the contacts should continue beneath the water.

c) How young are the **youngest** rocks in Michigan? Where are they located in the outcrop pattern? The oldest?

d) Strike and dip symbols are not shown for any of the units. Assuming that all dips are relatively gentle and that nothing has been overturned completely, use a black pen to indicate the strike and dip of the major units on the southern peninsula.

e) The outcrop pattern is similar to that associated with which kind of fold? How is this pattern different from that fold pattern? [HINT: Try to draw arrows to indicate the forces responsible for this structure.]

f) What are the possible ages of the deformation responsible for this outcrop pattern? Explain your evidence.

g) In the northwest corner of the map, Cambrian strata truncate several different kinds of Precambrian metamorphic and igneous rock; yet this contact is not mapped as a fault. What can it be? What does it represent?

h) Sketch a geologic cross section from east to west across the lower peninsula of Michigan.

i) A northwest-trending fault is shown just east of Lansing.

Assume that this is a strike-slip fault: describe the movement of the two blocks.

Now assume that it is a dip-slip fault: based on your interpretation of the dip of the strata, which side moved up and which moved down? Explain.

Examine figure 13.8, a geologic map showing the Black Hills of South Dakota and surrounding area.

j) Compare the outcrop pattern of the Black Hills with that of Michigan. What are the similarities? The differences?

A thick blanket of Tertiary sediment separates the Black Hills from the Rocky Mountains to the west. This material was eroded from the Rockies after the deformation that created the mountain system, and also fills basins west of the core of the mountain range.

k) Sketch a geologic cross section from the east side of the Black Hills to the foothills of the Rocky Mountains.

l) Carefully examine the contact between the Tertiary sediments and the **youngest** rocks in the Black Hills structure. What kind of contact is this?

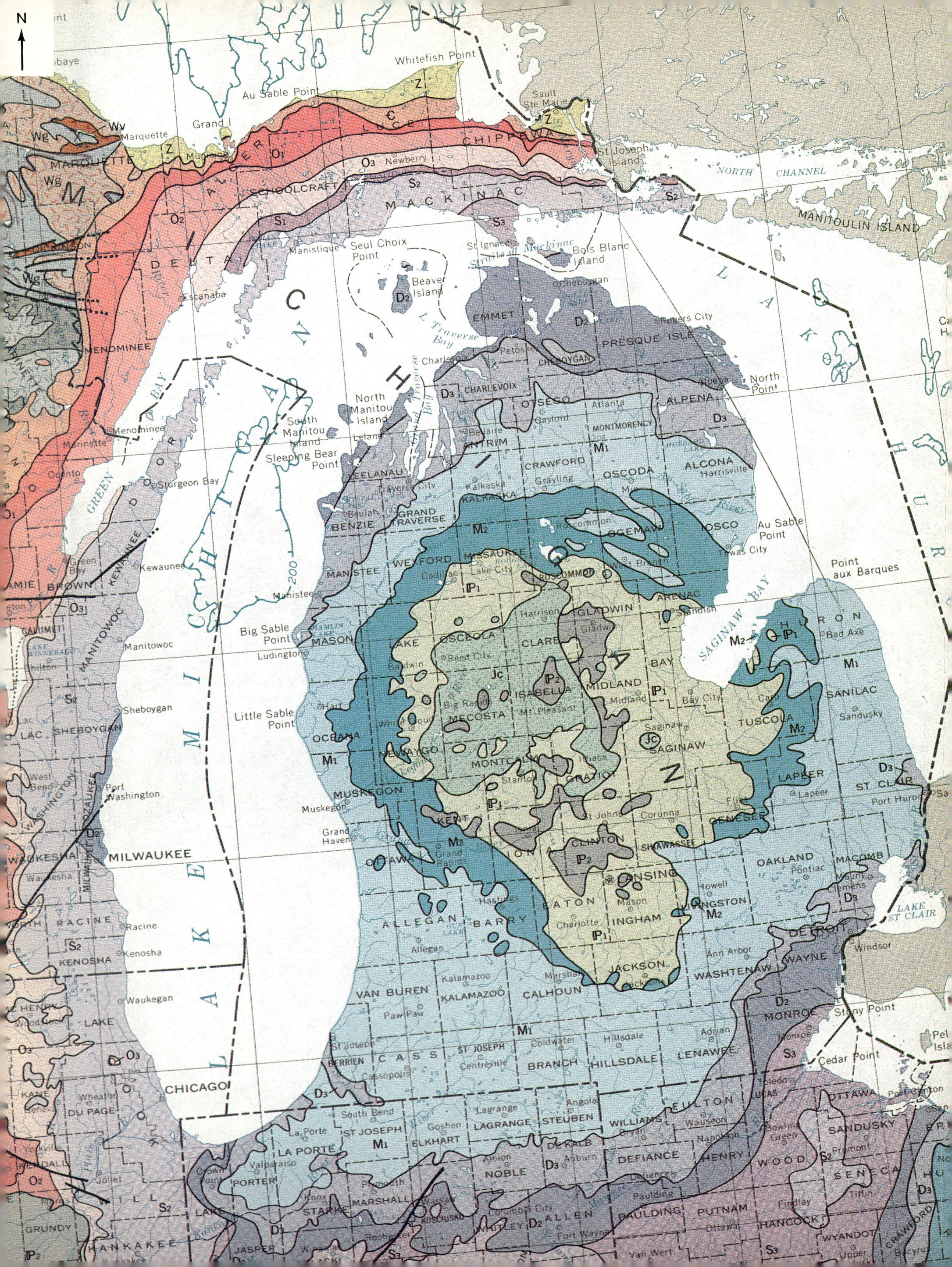

N
Whitefish Point
Au Sable Point
Grand I
Marquette
Munising
Sault Ste Marie
St Joseph Island
NORTH CHANNEL
MANITOULIN ISLAND
MARQUETTE
ALGER
LUCE
CHIPPEWA
SCHOOLCRAFT
MACKINAC
DELTA
MENOMINEE
Newberry
Manistique
Escanaba
Seul Choix Point
St Ignace
Straits of Mackinac
Bois Blanc Island
Cheboygan
Beaver Island
L Traverse Bay
EMMET
PRESQUE ISLE
CHEBOYGAN
Rogers City
Petoskey
Charlevoix
CHARLEVOIX
OTSEGO
Atlanta
ALPENA
Alpena
North Point
LAKE MICHIGAN
LAKE HURON
North Manitou Island
South Manitou Island
Sleeping Bear Point
Leland
ANTRIM
Bellaire
Gaylord
MONTMORENCY
LEELANAU
Traverse City
CRAWFORD
Grayling
OSCODA
Mio
ALCONA
Harrisville
KALKASKA
Kalkaska
BENZIE
Beulah
GRAND TRAVERSE
Roscommon
OGEMAW
IOSCO
Au Sable Point
Tawas City
West Branch
MANISTEE
WEXFORD
Cadillac
MISSAUKEE
Lake City
ROSCOMMON
ARENAC
Standish
Point aux Barques
Manistee
Big Sable Point
Ludington
MASON
LAKE
Baldwin
OSCEOLA
Reed City
Harrison
CLARE
Clare
GLADWIN
Gladwin
BAY
SAGINAW BAY
HURON
Bad Axe
ISABELLA
MIDLAND
Midland
Bay City
Caro
TUSCOLA
SANILAC
Sandusky
Little Sable Point
Hart
OCEANA
Big Rapids
MECOSTA
Mt Pleasant
Saginaw
SAGINAW
NEWAYGO
White Cloud
MONTCALM
Stanton
GRATIOT
Ithaca
LAPEER
Lapeer
ST CLAIR
Port Huron
MUSKEGON
Muskegon
KENT
St Johns
Corunna
Flint
GENESEE
Grand Haven
OTTAWA
Grand Rapids
IONIA
CLINTON
SHIAWASSEE
LANSING
OAKLAND
Pontiac
MACOMB
Mount Clemens
Howell
LIVINGSTON
Hastings
ALLEGAN
BARRY
EATON
Charlotte
Mason
INGHAM
Allegan
JACKSON
Jackson
Ann Arbor
WASHTENAW
DETROIT
WAYNE
LAKE ST CLAIR
Windsor
Kalamazoo
VAN BUREN
KALAMAZOO
Marshall
CALHOUN
Paw Paw
MONROE
Monroe
Stony Point
Cedar Point
St Joseph
BERRIEN
CASS
ST JOSEPH
Coldwater
BRANCH
Hillsdale
HILLSDALE
Adrian
LENAWEE
Cassopolis
Centreville
Toledo
LUCAS
OTTAWA
CHICAGO
South Bend
Goshen
ELKHART
Lagrange
LAGRANGE
Angola
STEUBEN
WILLIAMS
Bryan
FULTON
Wauseon
Napoleon
HENRY
Bowling Green
WOOD
SANDUSKY
Fremont
La Porte
LA PORTE
Valparaiso
PORTER
Albion
NOBLE
DE KALB
Auburn
DEFIANCE
Defiance
SENECA
Tiffin
Plymouth
MARSHALL
Warsaw
KOSCIUSKO
Columbia City
WHITLEY
ALLEN
Fort Wayne
Paulding
PAULDING
PUTNAM
Ottawa
Findlay
HANCOCK
WYANDOT
Upper
CRAWFORD
Bucyrus
Van Wert
JASPER
Rochester
KANKAKEE
GRUNDY
Joliet
KENDALL
Yorkville
KANE
Geneva
Wheaton
DU PAGE
COOK
LAKE
Waukegan
Kenosha
KENOSHA
RACINE
Racine
MILWAUKEE
Waukesha
WAUKESHA
WASHINGTON
West Bend
OZAUKEE
Port Washington
SHEBOYGAN
Sheboygan
MANITOWOC
Manitowoc
CALUMET
Chilton
BROWN
Green Bay
KEWAUNEE
Kewaunee
DOOR
Sturgeon Bay
GREEN BAY
Oconto
Marinette
Menominee
Z
Є
O1
O2
O3
S1
S2
S3
D2
D3
M1
M2
P1
P2
Jc
Wg
Wv
X
200

D3

Upper Devonian

D2

Middle Devonian

D1

Lower Devonian

S3

Upper Silurian
(Cayugan)

S2

Middle Silurian
(Niagaran)

S1

Lower Silurian
(Alexandrian)

O3

Upper Ordovician
(Cincinnatian)

O2

Middle Ordovician
(Mohawkian)
Upper Ordovician included in places

O1 | O1b | O1a

Lower Ordovician
(Canadian)
In Ozark Plateau of Missouri, divided into O1a *and* b *at base of Jefferson City Dolomite*

Eastern United States

Ꞓ | Ꞓq

Cambrian

Ꞓq, *basal Lower Cambrian clastic rocks, Chilhowee Group and equivalents, shown separately in parts of Appalachian region*

IP2

Des Moinesian Series
(= Allegheny Group)

IP1 | IP1a

Atokan and Morrowan Series
(=Pottsville Group)

P1a, *Jackfork Sandstone of Morrowan age in Ouachita Mountains*

M3

Chesterian Series

M2

Meramecian Series

M1

Osagean and Kinderhookian Series

Central interior region, from Appalachians to Kansas and central Texas

Wv

W volcanic rocks

Ely Greenstone and other units in Lake Superior region

Wg

W granitic rocks

2500 my and older

Ya

Anorthosite

Adirondack Mountains and elsewhere

X

X metasedimentary rocks

Includes (among other units): In Lake Superior region, Animikie Group and Marquette Range Supergroup; in South Dakota and Wyoming units in Black Hills and Medicine Bow Mountains; in Arizona, Mazatzal Quartzite, Yavapai Series, Pinal Schist, and metavolcanic formations. Iron formations of Lake Superior region separately shown in red

Y

Y sedimentary rocks

Includes (among other units): In Lake Superior region, Oronto Group of Keweenawan Supergroup, and Sioux Quartzite; in northwestern United States, undivided Belt Supergroup; in Utah, Big Cottonwood Group; in southwestern United States, Pahrump Group, Unkar Group of Grand Canyon Supergroup, Apache Group and Troy Quartzite; in northern Adirondack Mountains, Grenville Group

Z

Z sedimentary rocks

Includes (among other units): In Appalachian region, Ocoee Supergroup and Lynchburg Formation, including rocks older than 800 my; in Lake Superior region, Bayfield Group, Jacobsville and Hinckley Sandstones; in northwestern United States, Windermere Group, in Great Basin, Noonday, Johnnie, and other formations; in Grand Canyon, Chuar Group of Grand Canyon Supergroup

Yv

Y volcanic rocks

Mafic lavas in Keweenawan Supergroup. Felsic volcanic rocks in Missouri

Jc

Jurassic continental deposits

Minnesota, Iowa, and Michigan

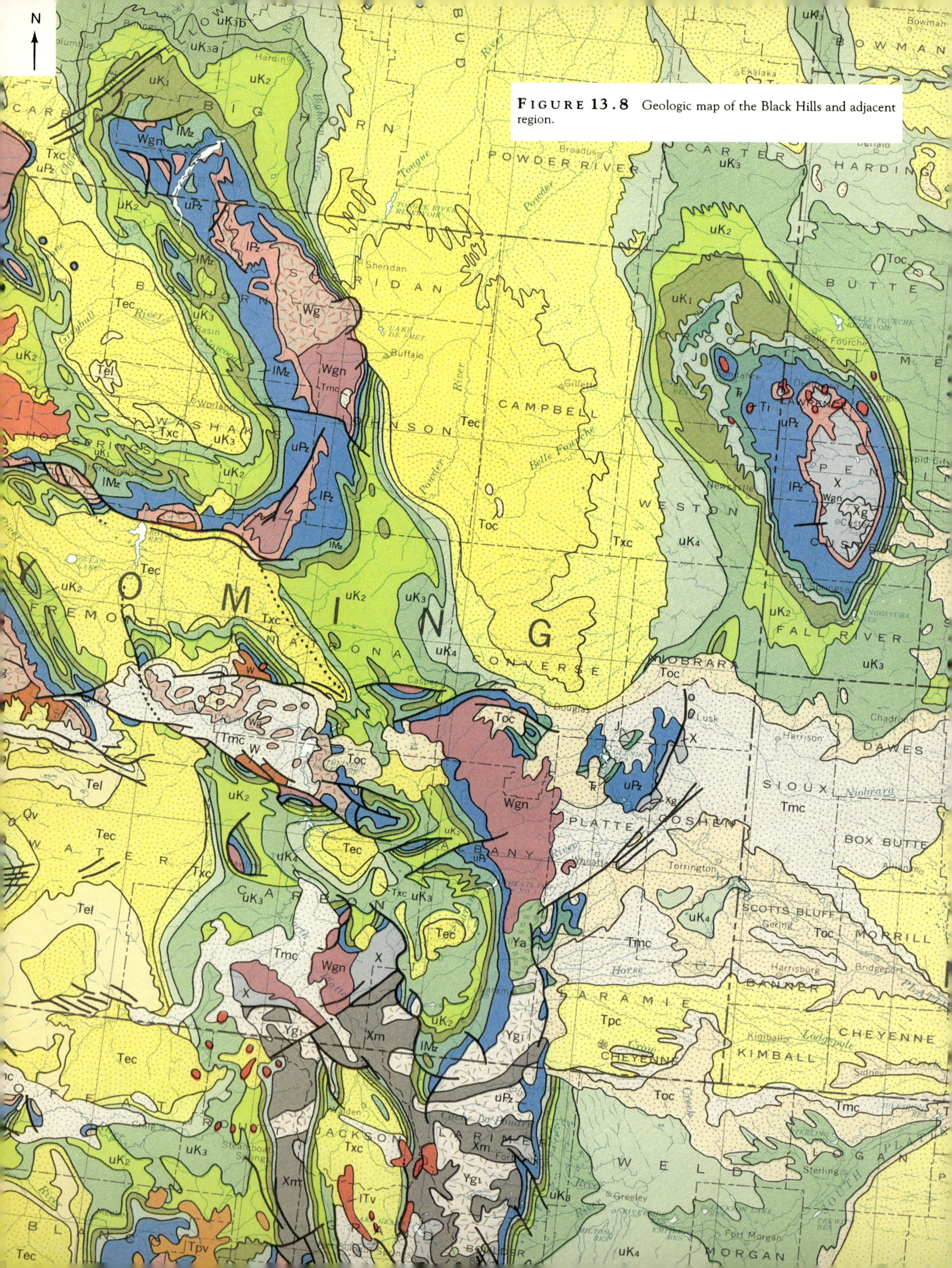

FIGURE 13.8 Geologic map of the Black Hills and adjacent region.

Atlantic and Gulf Coastal Plains and part of Great Plains

Qp

Pleistocene

Glacial deposits not included, see below

Atlantic and Gulf Coastal Plains and western interior region

uK_4

Navarro Group

(=Laramie and other formations)

uK_3 uK_3b uK_3a

Taylor Group

(=Montana Group)

In Montana plains, separated into uK_3a *and* b *at base of Judith River Formation*

uK_2

Austin and Eagle Ford Groups

(=Colorado Group)

uK_1

Woodbine and Tuscaloosa Groups

(=Dakota Group)

Lower Cretaceous included in parts of Rocky Mountain and High Plains regions

T̶R

Triassic

Mainly marine, but includes continental Dockum Group in Texas and New Mexico, and Newark Group in Appalachian region

Western United States and Piedmont province

lPz €

Lower Paleozoic

€, Cambrian separated in part

X

X metasedimentary rocks

Includes (among other units): In Lake Superior region, Animikie Group and Marquette Range Supergroup; in South Dakota and Wyoming units in Black Hills and Medicine Bow Mountains; in Arizona, Mazatzal Quartzite, Yavapai Series, Pinal Schist, and metavolcanic formations. Iron formations of Lake Superior region separately shown in red

Western United States

uPz P uPzc

Upper Paleozoic

P, *Permian separated in part;*
uPzc, *clastic wedge facies in Nevada and Idaho; Diamond Peak, Wood River, Milligen and associated formations*

Y

Y sedimentary rocks

Includes (among other units): In Lake Superior region, Oronto Group of Keweenawan Supergroup, and Sioux Quartzite; in northwestern United States, undivided Belt Supergroup; in Utah, Big Cottonwood Group; in southwestern United States, Pahrump Group, Unkar Group of Grand Canyon Supergroup, Apache Group and Troy Quartzite; in northern Adirondack Mountains, Grenville Group

J

Jurassic

Tmc

Miocene continental deposits

Toc

Oligocene continental deposits

Tec

Eocene continental deposits

Txc

Paleocene continental deposits

Ti

Tertiary intrusive rocks

Plugs, stocks, and laccoliths of all Tertiary ages; some granite plutons of early to middle Tertiary age

Yg_2

Younger

Y granite rocks

About 1000 my

Yg_1

Older

Y granitic rocks

About 1300–1400 my

Xg

X granitic rocks

About 1700–1800 my

Wg

W granitic rocks

2500 my and older

Xm

Orthogneiss and paragneiss

Extensive in western and southwestern United States; of varied compositions and origins

Wgn

Orthogneiss and paragneiss

m) What are the constraints on the age of the deformation that created the Black Hills structure?

n) What can you say about the age of the deformation responsible for the Rocky Mountains?

The "bulls-eye" outcrop patterns of Michigan and the Black Hills are typical of structures called basins and domes. These structures commonly occur in the midst of otherwise relatively undeformed rocks, suggesting a piston-like vertical movement, rather than the lateral compression responsible for more typical folds.

o) Locate other domes and basins near the Black Hills, and mark them with either a **D** or **B.**

p) Why are such uplifts difficult to explain with the plate tectonic model? (HINT: Compare the nature of the forces responsible for folding and faulting with those apparently present here.)

Examine figure 13.9, a geologic map of part of the southwestern United States.

q) Describe this map pattern in your own words.

r) Where are the youngest rocks? The oldest?

s) What does this map pattern suggest about the type of geologic structures present?

t) What are the constraints on the ages of those structures?

Examine figure 13.10, a simplified geologic map of Maine. Note the abundance of intrusive igneous rocks, particularly in the southern part of the state.

u) Where are the youngest **sedimentary** (stratified) rocks? The oldest?

v) A series of folds in the center of the state is outlined by the contact between units shown in light gray and green. Based on information available on this map, when did that folding take place? Explain your evidence.

w) Draw large arrows to indicate the direction of the compressional forces responsible for that folding.

x) The St. John and Allagash rivers in the northern part of the state flow roughly parallel to the axis of a large fold. Based on the ages of the rocks flanking these rivers, what kind of fold is it? Explain your reasoning.

y) A major fault zone cuts through southern Maine roughly parallel to the coast. What evidence can you cite that provides information concerning the age of faulting?

FIGURE 13.9 Geologic map of part of the southwestern United States.

Q

Quaternary
Only large areas and thick deposits shown

Western United States

uK

Upper Cretaceous

lK

Lower Cretaceous

lMz

Lower Mesozoic

P

Permian
Smaller areas included in undifferentiated upper Paleozoic

Colorado Plateau and Great Basin

lPz | €

Lower Paleozoic
€, Cambrian separated in part

Western United States and Piedmont province

Tpc

Pliocene continental deposits
Includes Miocene and Quaternary deposits in places

Tec

Eocene continental deposits

uMze

Upper Mesozoic eugeosynclinal deposits
Lower Cretaceous to Upper Jurassic, but with uncertain limits; includes Franciscan assemblage of California

Western California, Oregon and Washington

lMze

Lower Mesozoic eugeosynclinal deposits
Middle and lower Jurassic and Triassic

Western United States

TRPe

Triassic and Permian eugeosynclinal deposits
Dominantly volcanic

Oregon, California and Nevada

lPze

Lower Paleozoic eugeosynclinal deposits

Western United States

Qv | Qf

Quaternary volcanic rocks
Qf, *Felsic volcanic rock*

Tpv | Tpf

Pliocene volcanic rocks
Tpf, *felsic volcanic rocks*

Tmv | Tmf

Miocene volcanic rocks
Tmf, *Felsic volcanic rocks*

lTv | lTf

Lower Tertiary volcanic rocks
Oligocene and Eocene; some Cretaceous included in Arizona and elsewhere
lTf, *felsic volcanic rocks in southwestern Colorado and New Mexico*

Teb

Eocene marine pillow basalt

Coast Ranges of Washington and Oregon

Ti

Tertiary intrusive rocks
Plugs, stocks, and laccoliths of all Tertiary ages; some granite plutons of early to middle Tertiary age

Kg_2

Upper Cretaceous granitic rocks
80–90 my

Kg_1

Lower Cretaceous granitic rocks
105–120 my

Sierra Nevada, California

Jg

Jurassic granitic rocks

TRg

Triassic granitic rocks

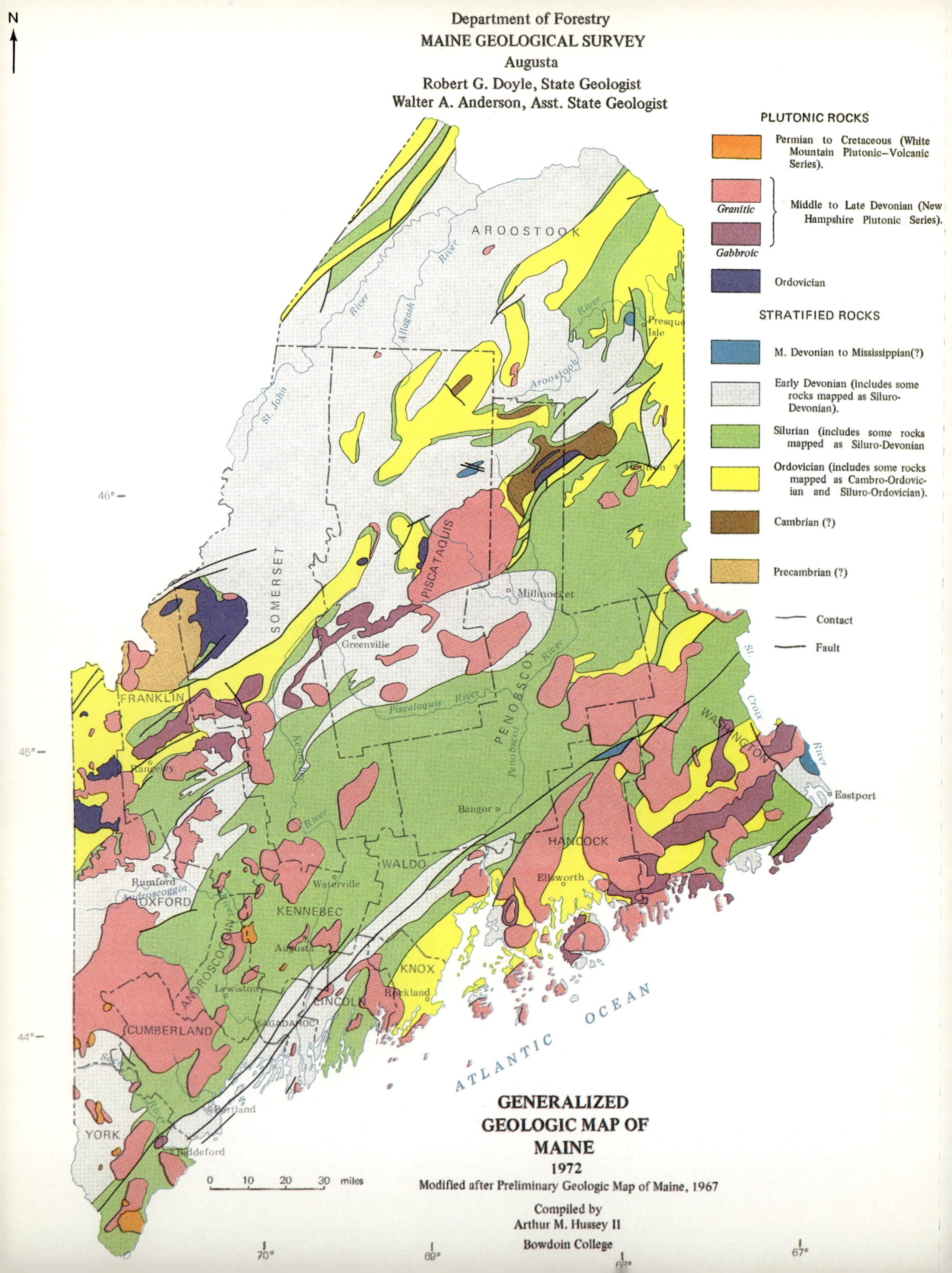
N
Department of Forestry
MAINE GEOLOGICAL SURVEY
Augusta
Robert G. Doyle, State Geologist
Walter A. Anderson, Asst. State Geologist
PLUTONIC ROCKS
Permian to Cretaceous (White Mountain Plutonic–Volcanic Series).
Granitic
Gabbroic
Middle to Late Devonian (New Hampshire Plutonic Series).
Ordovician
STRATIFIED ROCKS
M. Devonian to Mississippian(?)
Early Devonian (includes some rocks mapped as Siluro-Devonian).
Silurian (includes some rocks mapped as Siluro-Devonian
Ordovician (includes some rocks mapped as Cambro-Ordovician and Siluro-Ordovician).
Cambrian (?)
Precambrian (?)
Contact
Fault
AROOSTOOK
River
Allagash
River
St. John
Presque Isle
Aroostook
46°
SOMERSET
PISCATAQUIS
Millinocket
Greenville
FRANKLIN
Piscataquis River
PENOBSCOT
Penobscot
River
WASHINGTON
St. Croix River
45°
Rangeley
Bangor
Eastport
HANCOCK
WALDO
Rumford
Androscoggin
OXFORD
Waterville
Ellsworth
KENNEBEC
Augusta
KNOX
Lewiston
ANDROSCOGGIN
Rockland
LINCOLN
SAGADAHOC
CUMBERLAND
44°
ATLANTIC OCEAN
Portland
YORK
Biddeford
0 10 20 30 miles
GENERALIZED GEOLOGIC MAP OF MAINE
1972
Modified after Preliminary Geologic Map of Maine, 1967
Compiled by
Arthur M. Hussey II
Bowdoin College
70°
69°
67°

EXERCISE 13.3 A PAIR OF CHALLENGES TO YOUR ABILITIES AS GEOLOGICAL DETECTIVES

The following problems are presented with minimal help. Use what you have learned to describe the geologic histories of these two areas in as much detail as you can, citing geologic time of rock deposition and deformational events.

Figure 13.11 shows the southern end of the Appalachian Mountain system where it comes into contact with the Gulf Coastal Plain. Three different types of map patterns are present; essentially the northeastern, northwestern, and southern parts of the map.

a) Determine the geologic history within each of the three areas characterized by a different map pattern.

b) Describe the relations among the three different areas. How did these relations evolve?

c) How much geologic time is represented by the rocks of this map area?

FIGURE 13.10 (Preceding page). Generalized geologic map of Maine. Modified after Preliminary Geologic Map of Maine, 1967. Compiled by Arthur M. Hussey II, Bowdoin College.

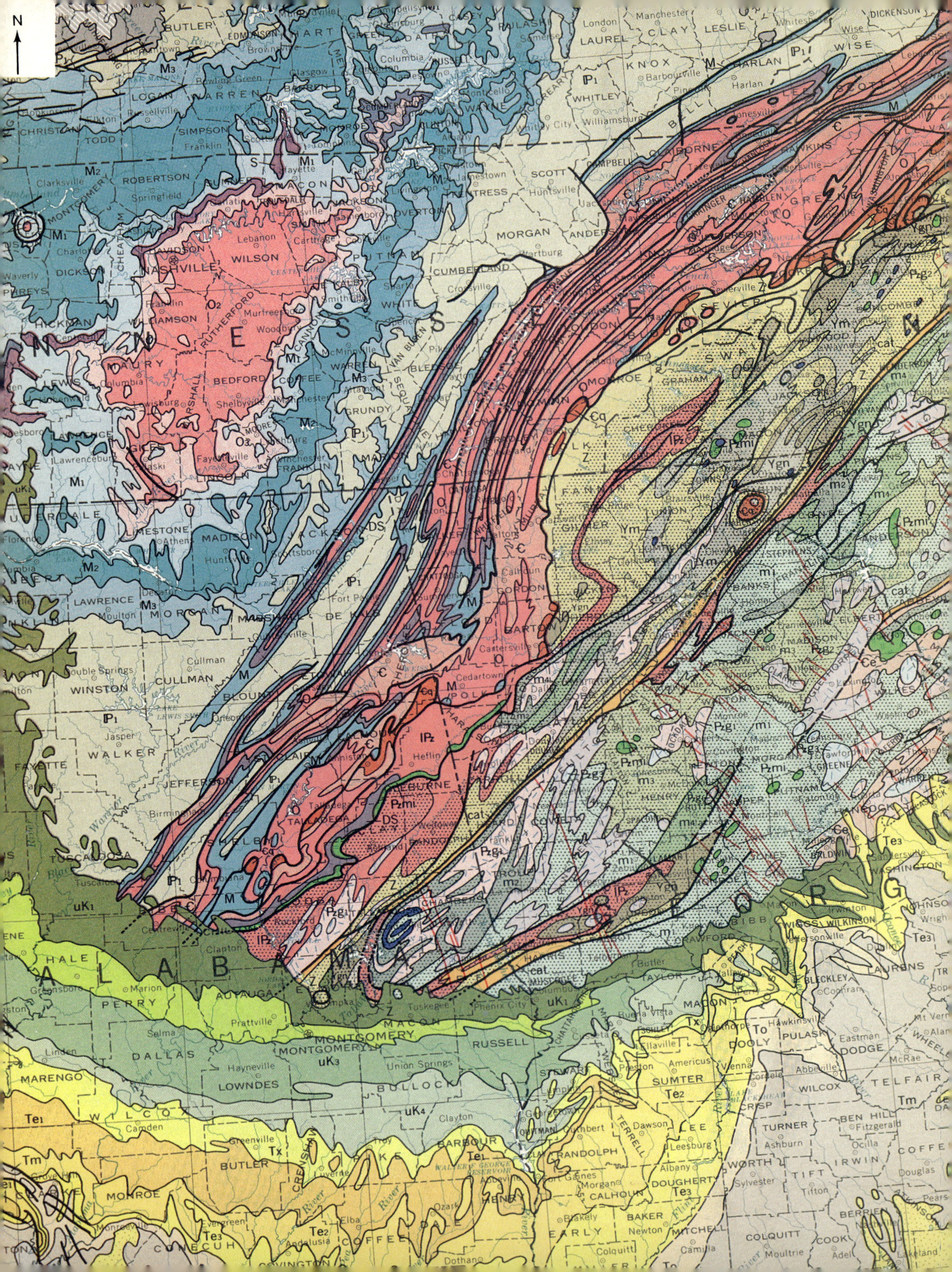

N
T E N N E S S E E
A L A B A M A
G E O R G I A
NASHVILLE
MONTGOMERY
BUTLER
LOGAN
WARREN
SIMPSON
CHRISTIAN
TODD
ROBERTSON
MONTGOMERY
DICKSON
HICKMAN
WILSON
RUTHERFORD
BEDFORD
COFFEE
GRUNDY
FRANKLIN
MADISON
MORGAN
LAWRENCE
CULLMAN
WINSTON
WALKER
FAYETTE
JEFFERSON
TUSCALOOSA
HALE
PERRY
AUTAUGA
DALLAS
LOWNDES
MARENGO
MONROE
BUTLER
BULLOCK
RUSSELL
BARBOUR
COFFEE
CONECUH
COVINGTON
CUMBERLAND
SCOTT
MORGAN
KNOX
LAUREL
CLAY
LESLIE
HARLAN
WHITLEY
WISE
SUMTER
DOOLY
PULASKI
DODGE
TELFAIR
WILCOX
TURNER
BEN HILL
IRWIN
TIFT
WORTH
LEE
DOUGHERTY
CALHOUN
BAKER
MITCHELL
COLQUITT
COOK
BERRIEN
EARLY
RANDOLPH
TERRELL
MACON
TAYLOR
BIBB
WILKINSON
WASHINGTON
Bowling Green
Franklin
Springfield
Clarksville
Lebanon
Murfreesboro
Columbia
Shelbyville
Lewisburg
Fayetteville
Winchester
Athens
Huntsville
Scottsboro
Fort Payne
Cullman
Jasper
Birmingham
Talladega
Heflin
Cedartown
Cartersville
Calhoun
Tuscaloosa
Centreville
Clanton
Prattville
Selma
Marion
Greensboro
Hayneville
Union Springs
Tuskegee
Phenix City
Troy
Clayton
Greenville
Camden
Elba
Ozark
Dothan
Andalusia
Evergreen
Americus
Dawson
Albany
Leesburg
Camilla
Moultrie
Sylvester
Ashburn
Fitzgerald
Ocilla
Douglas
Hawkinsville
Eastman
McRae
Cochran
Abbeville
Vienna
Cordele
Butler
Buena Vista
Harlan
Barbourville
Williamsburg
Huntsville
Wartburg
Crossville
Jamestown
M₁
M₂
M₃
O₂
O₃
uK₁
uK₃
uK₄
Te₁
Te₂
Te₃
Tm
Tx
To
Ym
Ygn
cat
m₁
m₂
m₃
m₄
Pzmi
Pzg
DS

Tm

Miocene

Pliocene included in places

Includes Pliocene in Southeastern U.S.

To

Oligocene

Te3

Jackson Group

Te2

Claiborne Group

Te1

Wilcox Group

Atlantic and Gulf Coastal Plains

Tx

Paleocene

Includes Midway Group of Gulf Coastal Plain

uK4

Navarro Group
(=Laramie and other formations)

uK3 | uK3b | uK3a

Taylor Group
(=Montana Group)

In Montana plains, separated into uK3a *and* b *at base of Judith River Formation*

uK2

Austin and Eagle Ford Groups
(=Colorado Group)

uK1

Woodbine and Tuscaloosa Groups
(=Dakota Group)

Lower Cretaceous included in parts of Rocky Mountain and High Plains regions

Atlantic and Gulf Coastal Plains and western interior region

cat

Cataclastic rocks

Produced by Paleozoic orogenies

Piedmont province

ℙ1 | ℙ1a

Atokan and Morrowan Series
(=Pottsville Group)

P1a, *Jackfork Sandstone of Morrowan age in Ouachita Mountains*

M3

Chesterian Series

M2

Meramecian Series

M1

Osagean and Kinderhookian Series

Central interior region, from Appalachians to Kansas and central Texas

O3

Upper Ordovician
(Cincinnatian)

O2

Middle Ordovician
(Mohawkian)

Upper Ordovician included in places

Eastern United States

€ | €q

Cambrian

€q, *basal Lower Cambrian clastic rocks, Chilhowee Group and equivalents, shown separately in parts of Appalachian region*

Z

Z sedimentary rocks

Includes (among other units): In Appalachian region, Ocoee Supergroup and Lynchburg Formation, including rocks older than 800 my: in Lake Superior region, Bayfield Group, Jacobsville and Hinckley Sandstones; in northwestern United States, Windermere Group; in Great Basin, Noonday, Johnnie, and other formations; in Grand Canyon, Chuar Group of Grand Canyon Supergroup

D

Devonian

S

Silurian

ℙz | €

Lower Paleozoic

€, *Cambrian separated in part*

Western United States and Piedmont province

Pzg3

Upper Paleozoic granitic rocks

About 250–300 my. Mainly in southeast part of Piedmont province, where they are post-metamorphic; rare in New England

Pzg2

Middle Paleozoic granitic rocks

About 350–400 my. In New England, New Hampshire Plutonic Series and others, related to Acadian orogeny; in Piedmont province, synmetamorphic

Appalachian region (New England and Piedmont province), rarely in western United States

€e

Cambrian eugeosynclinal deposits

Pzmi

Paleozoic mafic intrusives

Of various ages, not differentiated

Appalachian region

ms | m1 | m2 | m3 | m4

Metamorphic complexes

Metamorphosed during Phanerozoic orogenies. Ages of original rocks uncertain; in Piedmont province may be largely Precambrian Z, but with some lower Paleozoic; in northern Washington may be largely lower to upper Paleozoic

ms, *schist and phyllite*

m_1, *felsic paragneiss and schist*

m_2, *mafic paragneiss (=hornblendite and amphibolite)*

m_3, *migmatite*

m_4, *felsic orthogneiss (=granite gneiss)*

Piedmont province and northern Washington

Ym

Paragneiss and schist

Ygn

Orthogneiss

Figure 13.12 is a photograph of part of the Moon. The same principles that we use to unravel geologic history on Earth can be applied to other planets and satellites as well. On the Moon, the major event is cratering produced by the impact of meteorites. Each impact carves a crater in the lunar surface; the excavated material is blasted upward by the shock, spreads out, and settles back to the surface. This process forms a deposit called an **ejecta blanket,** which is easily visible in many cases.

a) Select a prominent crater and label it **x.** Determine the relative ages of adjacent craters and ejecta blankets, and label them **x + 1, x + 2,** and **x + 3,** in appropriate sequence, if they are **younger** than x. Label them **x − 1, x − 2,** and **x − 3,** if they are **older** than crater x.

b) Piece together the history of this portion of the Moon as best you can. How many different events are definitely represented?

FIGURE 13.12 A photograph of the moon showing several crater events.

CHAPTER 14

PLATE TECTONICS

PURPOSE

To study some of the processes that are part of the plate tectonic model for Earth behavior, and to calculate the rates of some of those processes.

MATERIALS NEEDED

1) Tracing paper.
2) Colored pencils.
3) A ruler with close-spaced divisions (tenths of an inch or millimeters).
4) A protractor.
5) A calculator or pencil and paper for simple arithmetic.

INTRODUCTION

The modern view of the Earth is that it is an active, ever-changing planet, on which internal processes bring about major changes in the appearance and location of even the largest surface features—mountain systems, continents, ocean basins. The plate tectonic model, described in detail in your textbook, is our best hypothesis about the processes that bring about these changes. The plate tectonic model, illustrated in figures 14.1 and 14.2, can be summarized briefly as follows.

The Earth's rigid outermost layer (the **lithosphere**) is broken into a small number of fragments called **lithosphere plates.** Each is quite thin relative to the area it covers. These plates are able to move because they rest upon a zone of low rigidity (the **asthenosphere**), which can flow easily. Each plate can move independently, enabling it to have three kinds of interaction with neighboring plates. These interactions lead to three different kinds of boundaries between plates:

1) **Divergent plate boundaries** form where a plate is split, and the pieces move away from one another. These correspond to oceanic ridges (sea-floor spreading centers) and to major rift zones on the continents. Magma rises upward into the zone of rifting and creates new oceanic crust.
2) **Convergent plate boundaries** form where two plates collide head-on with one another. Oceanic crust that formed originally at spreading centers is returned to the mantle by being thrust downward in **subduction zones,** which correspond to ocean trenches. Partial melting of the subducted plate creates magmas, which rise to form **volcanic island arcs** in the oceans or **Andean-type mountains** on the continents.
3) **Transform-fault boundaries** form where two plates grind past one another without colliding head-on. These correspond to major strike-slip faults on the continents and to the fracture zones of the oceans.

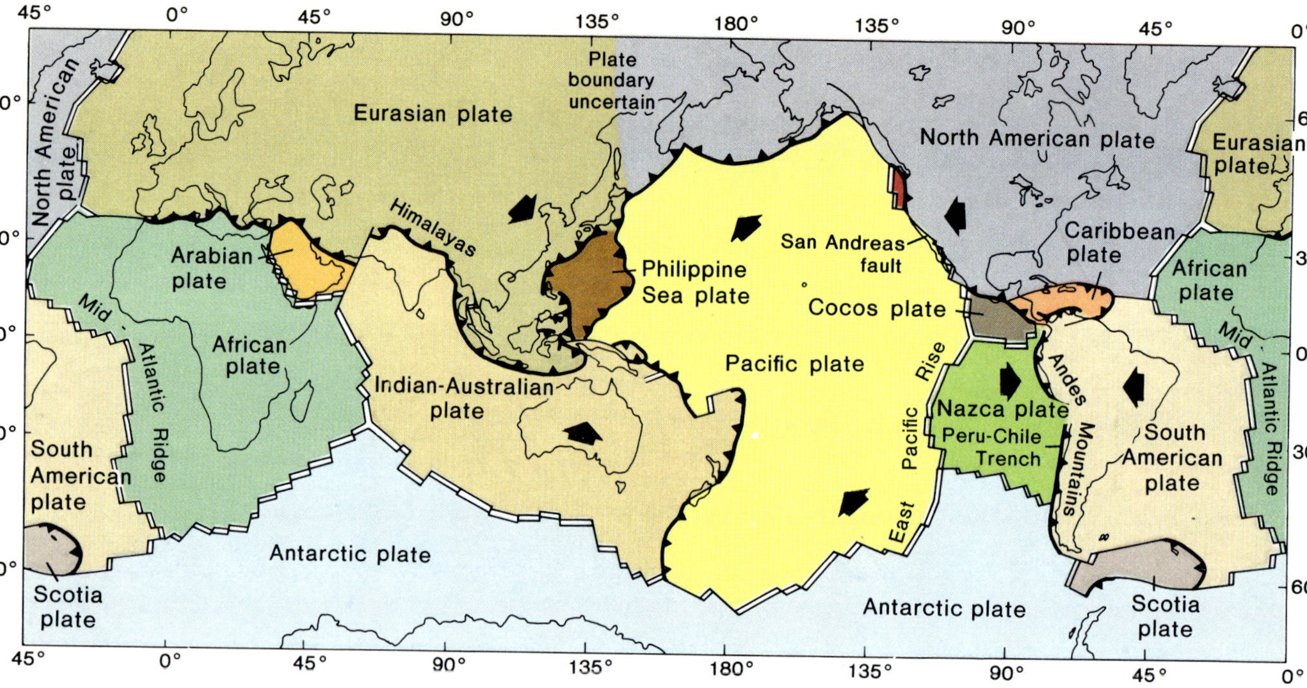

FIGURE 14.1 The major plates of the world. The western edge of the map repeats the eastern edge so that all plates can be shown unbroken. Double lines indicate spreading centers on diverging plate boundaries. Single lines show transform boundaries. Heavy lines with triangles show converging boundaries, with triangles pointing down subduction zones. Modified from W. Hamilton, U.S. Geologic Survey.

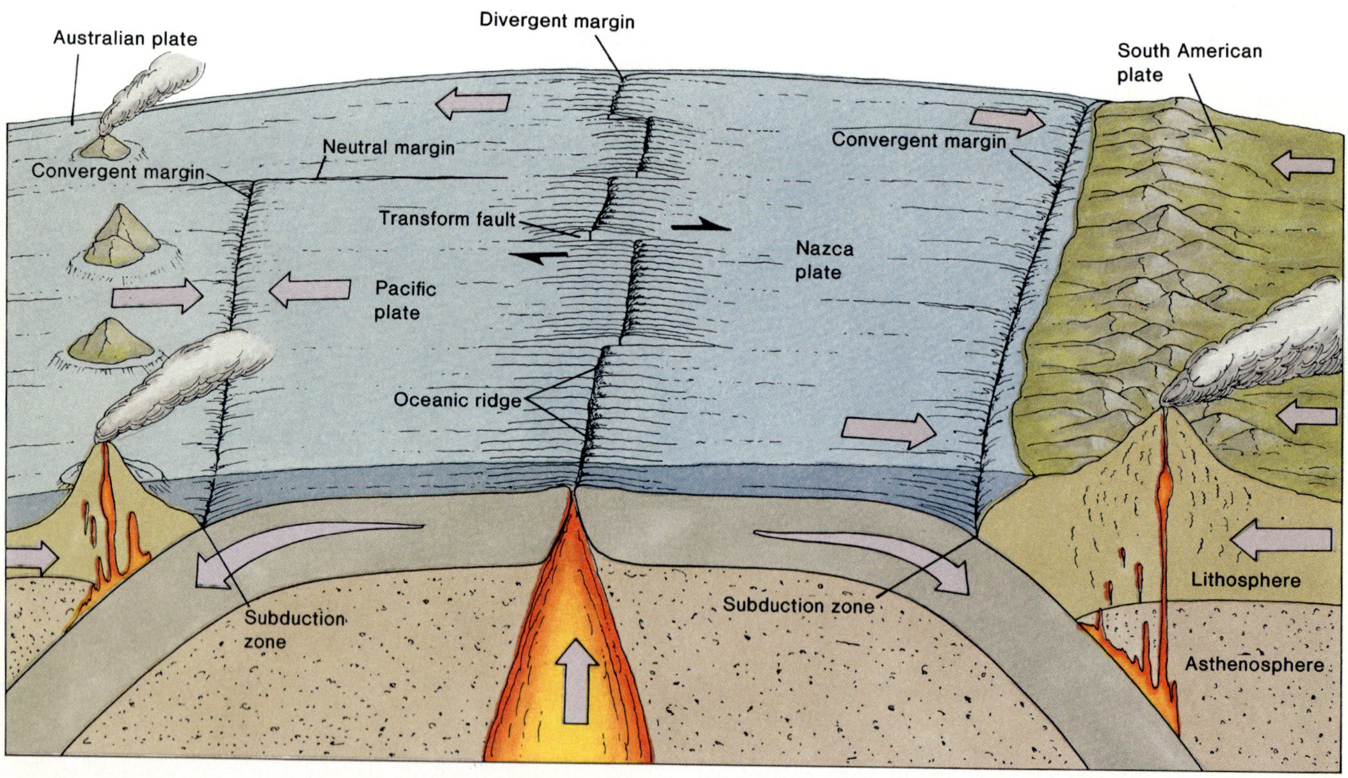

FIGURE 14.2 Cross-section showing plate boundaries at depth.

The plate tectonic model is viewed as the best explanation for Earth behavior by most geologists because it satisfactorily explains many features and processes, including the locations of volcanoes and earthquakes, the origins and geographic distributions of different kinds of rock types, mountains, metamorphism, fossils, and other phenomena. The exercises that follow use some of these aspects of the Earth to analyze the nature and rate of plate motion.

Remember that, large as they are, continents are simply passengers on lithosphere plates, changing position as the plates beneath them shift.

EXERCISE 14.1 FITTING THE PIECES TOGETHER

Figure 14.3 is a map of the South Atlantic Ocean showing the Mid-Atlantic Ridge and the continents of Africa and South America. As early as 250 years ago, scientists observed that the two continents could be fitted together, tucking the east coast of South America into the crook in the west coast of Africa. We now think that they were, indeed, part of a single continent (Gondwanaland) before there was an Atlantic Ocean. The problem is, how best to put the two pieces back together?

a) On two pieces of tracing paper, sketch the outlines of Africa and South America, **using the present shoreline** as the edge of the continent.

b) What factors, other than plate tectonic ones, would affect the shapes of the continents in this case?

c) Sketch the outlines of the continents again, this time using the edges of their **continental shelves.**

d) Bring the tracings you made in (a) together to produce the best possible fit of Africa and South America.

e) Do the same with the tracings you made in (c).

f) Which version of the continents produces the best fit?

g) What does this tell you about the locations of the edges of all the continents?

h) Some areas don't fit very well in either reconstruction. Locate these, and suggest an explanation for the misfit. (HINT: Remember that geology is four-dimensional. Not all features form at the same time.)

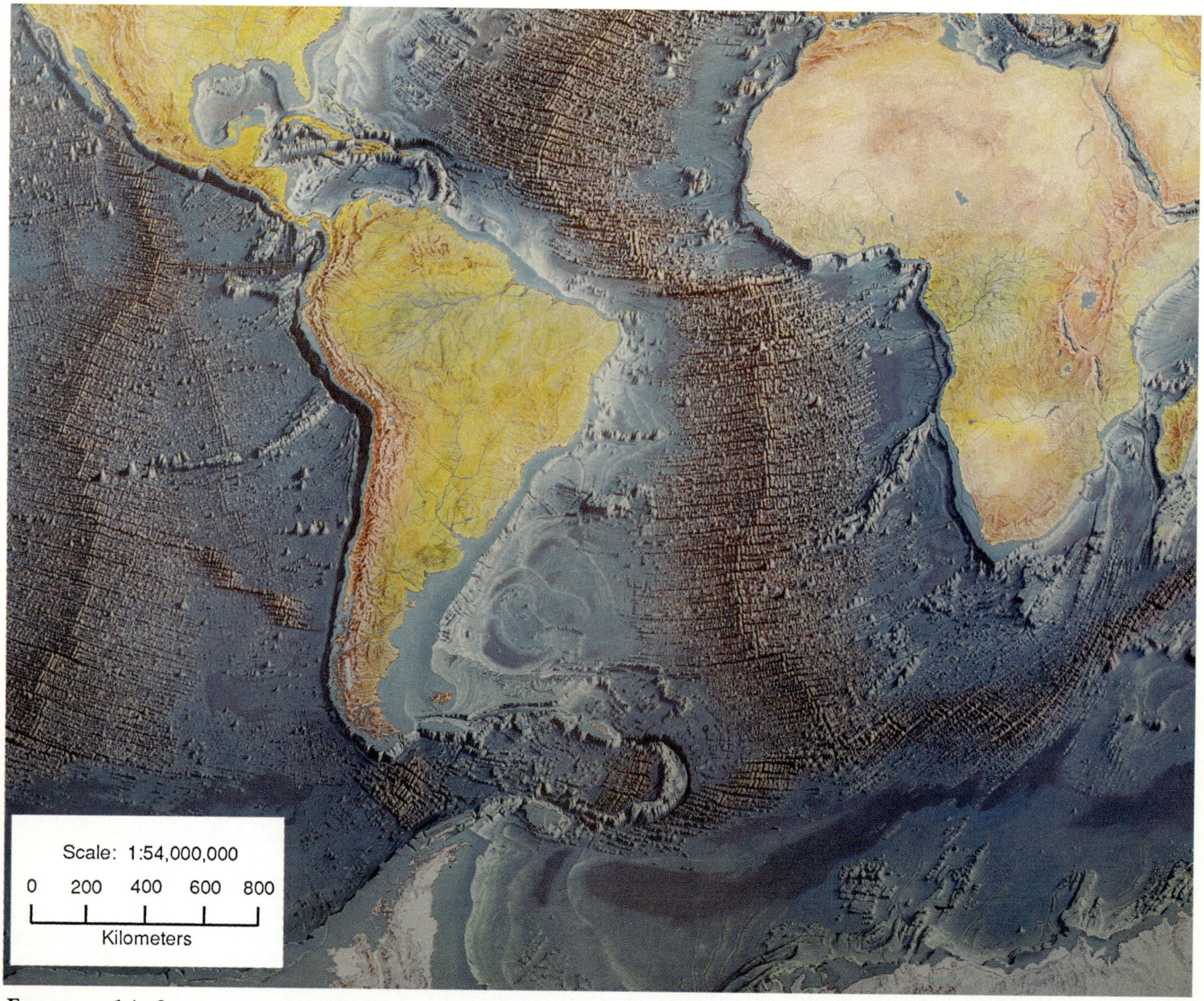

FIGURE 14.3 Physiographic map of the South Atlantic Ocean.

EXERCISE 14.2 DECIPHERING PLATE MOVEMENTS

Geologists use several techniques to determine former directions of plate motion. One method involves rocks that were deposited in ancient climates, which would be incompatible with their present geographic locations. For example, plants and animals now preserved as fossils in Spitsbergen, well above the Arctic Circle in Scandinavia, could have lived only in a temperate, mid- to low-latitude environment. Extensive ancient till deposits, now found in equatorial and near-equatorial regions of South America and Africa, could have resulted only from widespread continental glaciation at much higher latitudes. Both examples suggest plate motion across many degrees of latitude. Other methods utilize information about Earth's magnetic field that is now recorded in sedimentary and volcanic rocks.

One of the best methods for determining both the direction and the rate of plate motion is the study of oceanic island chains produced by melting of the upper mantle above **hotspots**—narrowly focused sources of heat. Figure 14.4 shows how such island chains form. Melting above the hotspot produces magma that erupts to form a volcano. If the plate on which the volcano rests then moves, the volcano is carried away from the heat source that created it. It becomes extinct, but a new volcano develops above the hotspot. As this process continues, a chain of volcanoes forms that marks the path of the moving plate above the hotspot, almost like footprints.

The Hawaiian Islands are one of the most famous chains of hotspot volcanoes. Figure 14.5 shows the Hawaiian Islands and gives the ages of the major volcanism on each island.

a) Where is the hotspot located today? How do you know?

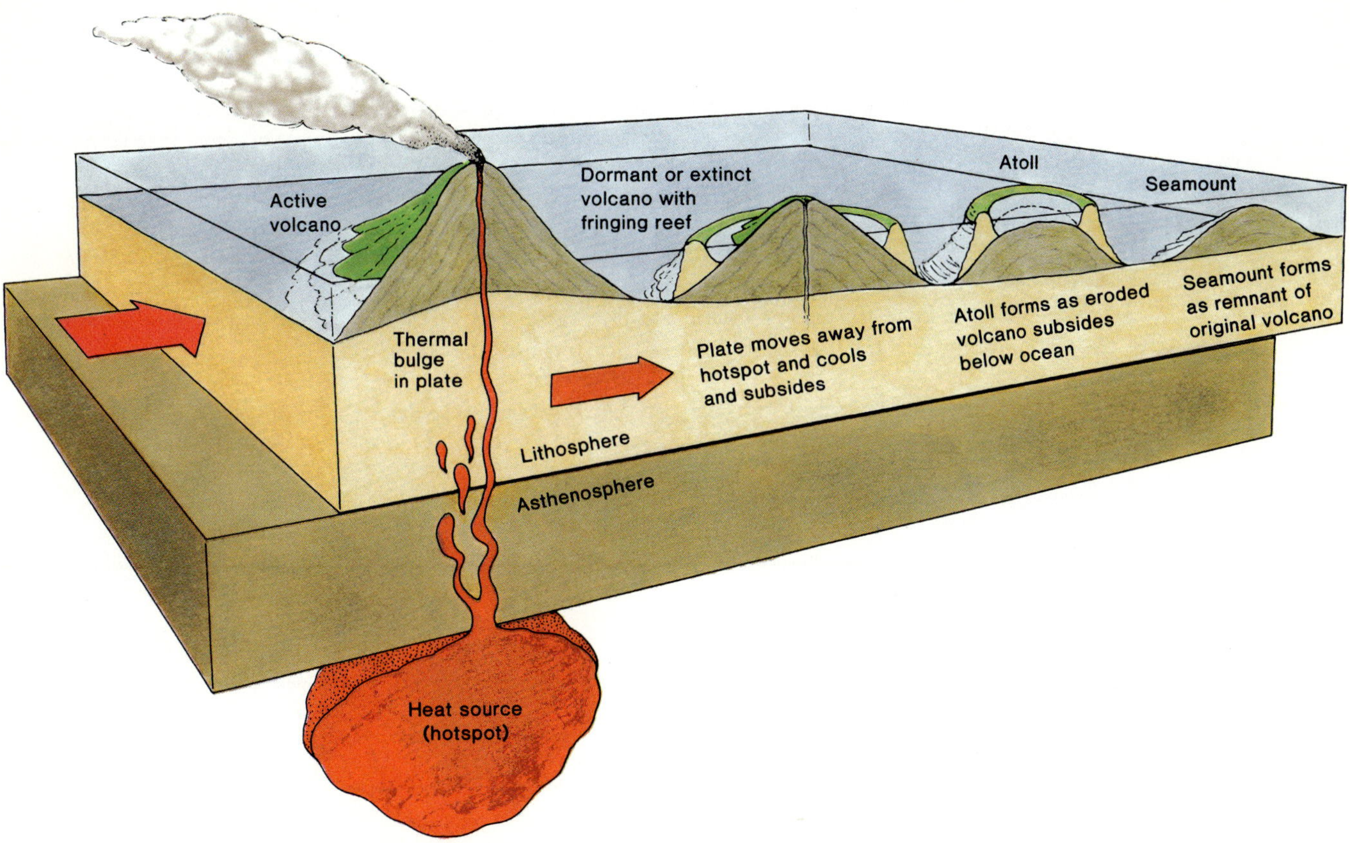

FIGURE 14.4 Development of Hawaiian-type volcanic island chains.

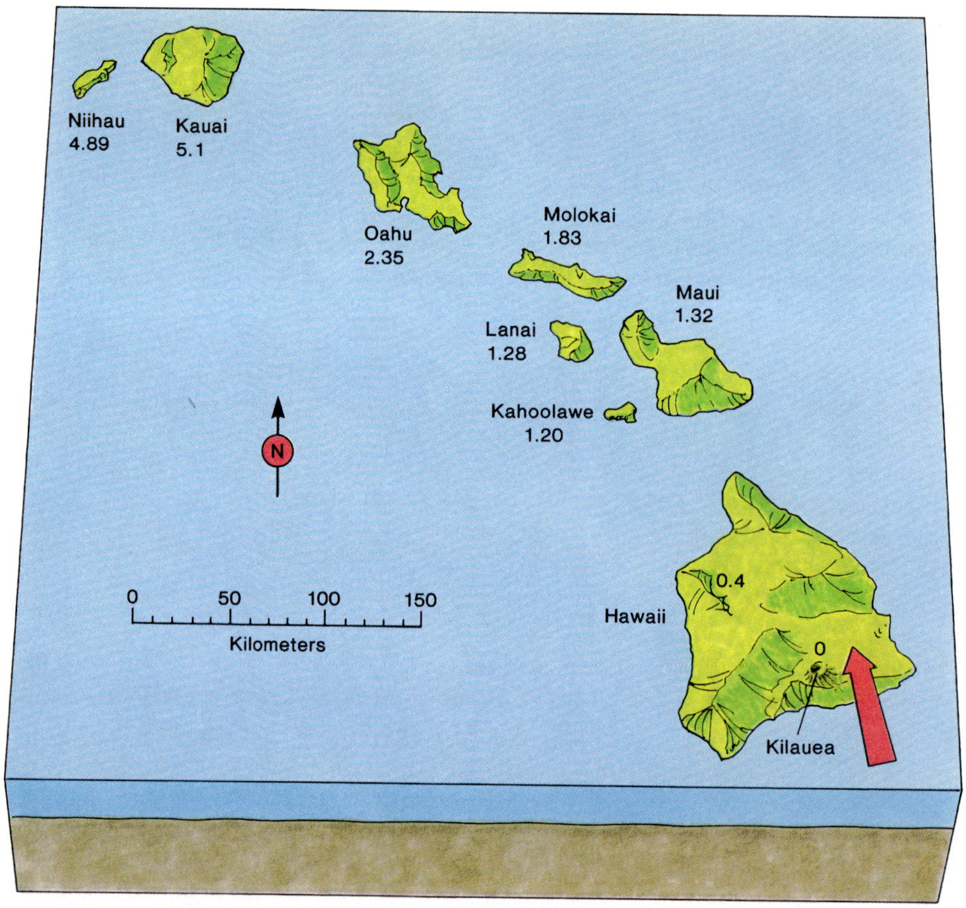

FIGURE 14.5 Ages of volcanism in the Hawaiian Islands (in millions of years).

b) Use a colored pencil to draw a line indicating the direction of motion as revealed by the positions of the major islands (Hawaii, Maui, Molokai, Oahu, Kauai). [Draw the line through the approximate center of each island, except for Hawaii itself; there, use the location of Kilauea crater.]

c) Use your ruler, protractor, and some simple arithmetic to fill in table 14.1. This will enable you to determine the directions and rates of plate motion for the Pacific plate beneath these islands.

TABLE 14.1 MOTION OF THE PACIFIC PLATE BENEATH THE HAWAIIAN ISLANDS

	DISTANCE BETWEEN ISLANDS	# YEARS OF PLATE MOTION	RATE OF PLATE MOTION (km/yr)	DIRECTION OF PLATE MOTION
Hawaii–Maui				
Maui–Molokai				
Molokai–Oahu				
Oahu–Kauai				

The Hawaiian Islands represent only the last 5 million years (m.y.) or so of movement by the Pacific plate, but a chain of submerged volcanic cones (seamounts) continues on to the north (Figure 14.6), indicating that the hotspot has been active for a much longer time. Some work with figure 14.6 will fit the most recent movement of the Pacific plate into its historic context.

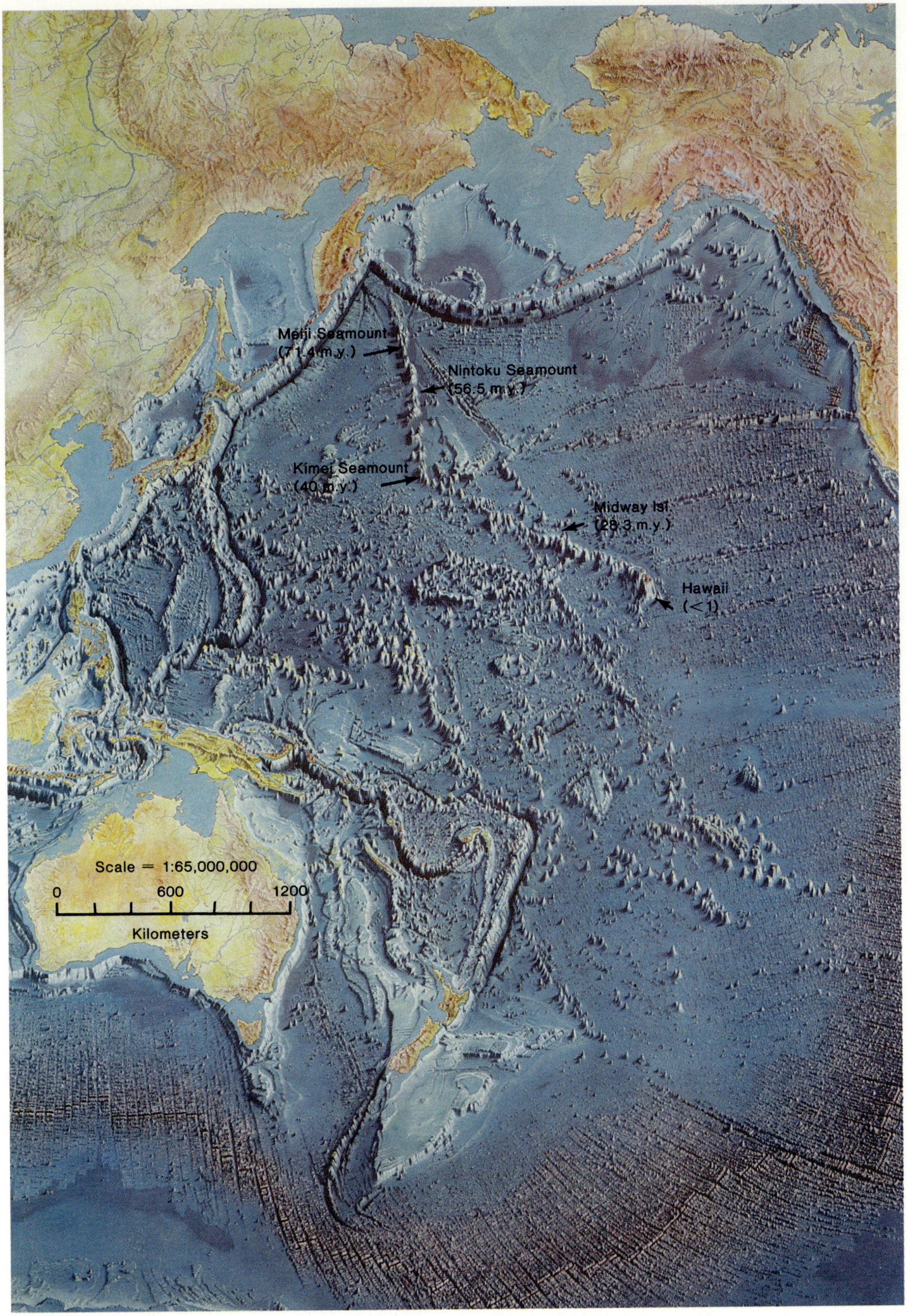

FIGURE 14.6 The floor of the Pacific Ocean showing ages of volcanoes in the Hawaiian-Emperor Seamount Chain.

d) What evidence is there that the Pacific plate has not always moved in a constant direction?

e) At approximately what point in Earth history did it change its direction? How did you determine this?

f) Based on the oldest data available in figure 14.6, in what direction did the Pacific plate originally move?

g) How far has the Meiji Seamount moved from the hotspot that created it? How did you determine this?

h) Has the Meiji Seamount always moved at approximately the same rate? Explain.

i) Assuming that the Pacific plate continues to move in its **present** direction, what will be the eventual fate of the Meiji Seamount?

j) Suggest an approximate time in the future when this fate will be met. How did you determine this?

Return to figure 14.3, the map of the South Atlantic. There is another way to roughly estimate the rate at which plates have been moving: by knowing when they started to move, and how long it took to get to their present positions.

k) Measure the distance from the easternmost point of South America to the westernmost point of Africa in the area just south of Lagos, Nigeria. Remember where South America and Africa actually end!! [To be most accurate, measure parallel to the fracture zones.]

l) Assuming that the oldest rock in the South Atlantic is of Jurassic age (approximately 185 m.y. old), determine the spreading rate of the South Atlantic Ocean in km/m.y. and cm/year. Explain how you determined this rate.

Sea-floor spreading rates can be measured much more accurately than this. Your textbook describes how Earth's magnetic field combines with the paleomagnetism recorded by ocean-floor basalts to produce magnetic anomaly "stripes" in the ocean basins. Each positive anomaly represents basalts that erupted at a mid-ocean ridge during a time when Earth's magnetic polarity was the same as it is today. Each negative anomaly represents basalts erupted at a ridge when magnetic polarity was reversed—the opposite of what it is today. By dating the basalts, we can determine the timing of the reversals and thus the rates of spreading.

Figure 14.7 contains magnetic maps of parts of three different oceanic ridges. Each is drawn at a different scale in order to keep **time** constant for all three, as shown by the time line at the top of the diagram. Black areas indicate positive anomalies, and white areas, negative anomalies. Note that these maps are only a part of **half** of each ocean basin; the ridge-crest spreading center is near the right side of each map. As a result, any spreading rate that you calculate will be **half** of the true rate.

m) Why is the calculated spreading rate only half the true rate?

n) Measure the total distance of sea-floor spreading from each ridge for the time interval indicated.

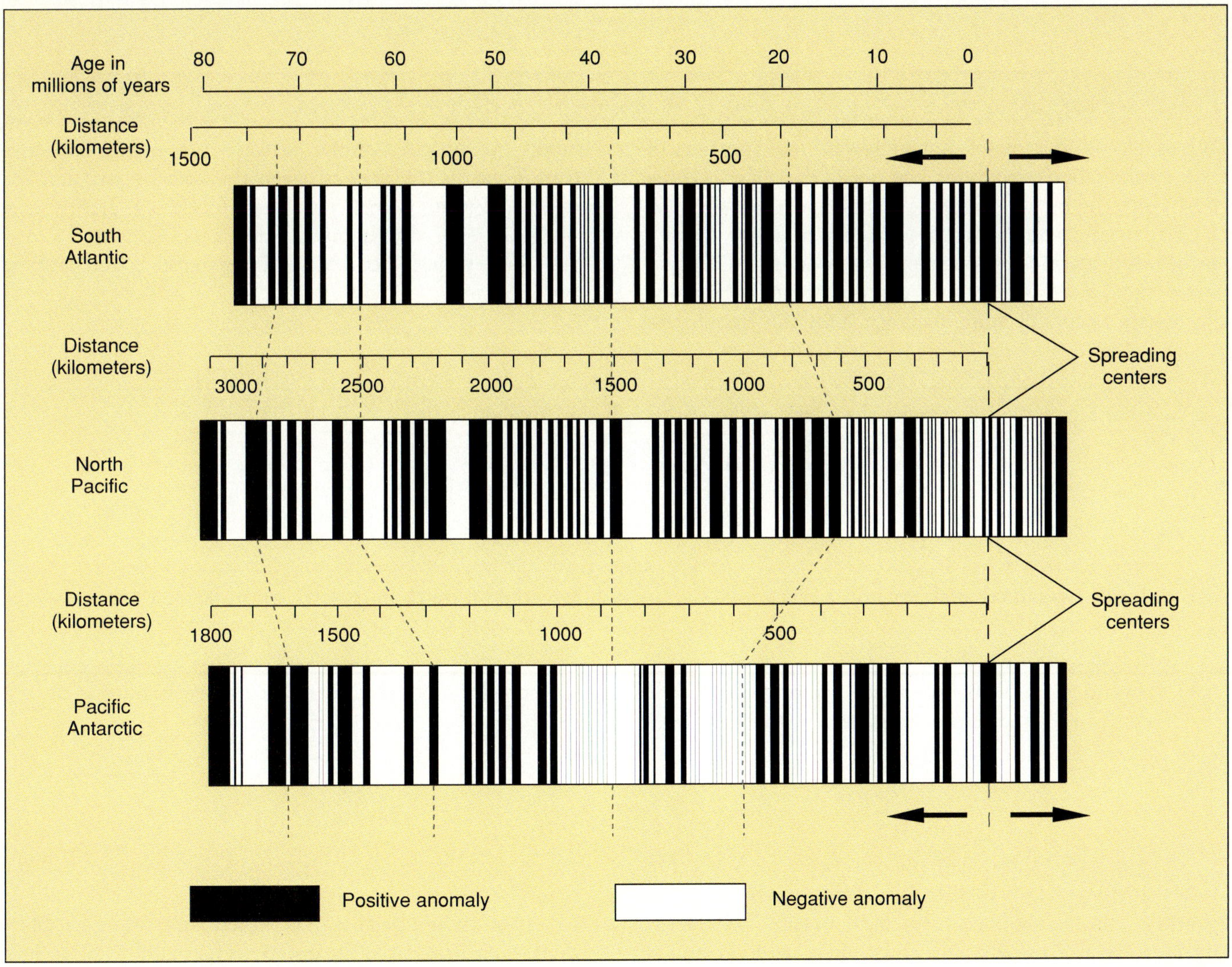

FIGURE 14.7 Magnetic anomaly profiles across three ocean ridges. Dotted lines connect same anomaly on each ridge.

o) Calculate the rate of spreading for each ridge (remember to multiply by 2).

p) How do the spreading rates calculated for the South Atlantic and North Pacific oceans compare with the rates of motion that you calculated in the previous exercises? What sources of error or other explanations might explain any discrepancies?

q) Examine the profile for the South Atlantic in 20-m.y. segments. Was the spreading rate constant or did it vary?

r) What does this result suggest about the accuracy of long-range predictions such as that made for the Meiji Seamount?

EXERCISE 14.3 ANATOMY OF SUBDUCTION ZONES

When two oceanic plates moving toward one another collide, a subduction zone forms, and one of the plates is thrust down into the mantle (Figure 14.8). The descending plate melts partially when it reaches a depth at which there is sufficient heat, and the magma rises upward to build a volcanic island arc. Because of the depth necessary for melting, volcanism occurs at some distance from the trench that marks the surface contact between two plates. The **arc-trench gap** is the area between the volcanic arc and the trench. Within it are commonly found an accretionary wedge, composed of highly deformed sediment, and a fore-arc basin, which accumulates debris eroded and erupted from the arc.

a) What factors will determine the width of the arc-trench gap? Explain.

b) Sketch a profile that shows how a much wider arc-trench gap could form than that shown in figure 14.8.

c) Explain the probable sources of sediment that would be found on the subducted plate (at the left hand side of figure 14.8) and in the marginal sea (to the right).

Figure 14.9 shows partially completed profiles across four segments of the Aleutian island arc, which extends to the west from mainland (continental) Alaska. The relative positions of the volcanic arc and trench are indicated in each profile. The foci of subduction-related earthquakes are indicated in figure 14.9a, defining a narrow region called a Benioff Zone.

You should be able to complete the profiles based on what you learned from the questions for figure 14.8. To help a little:

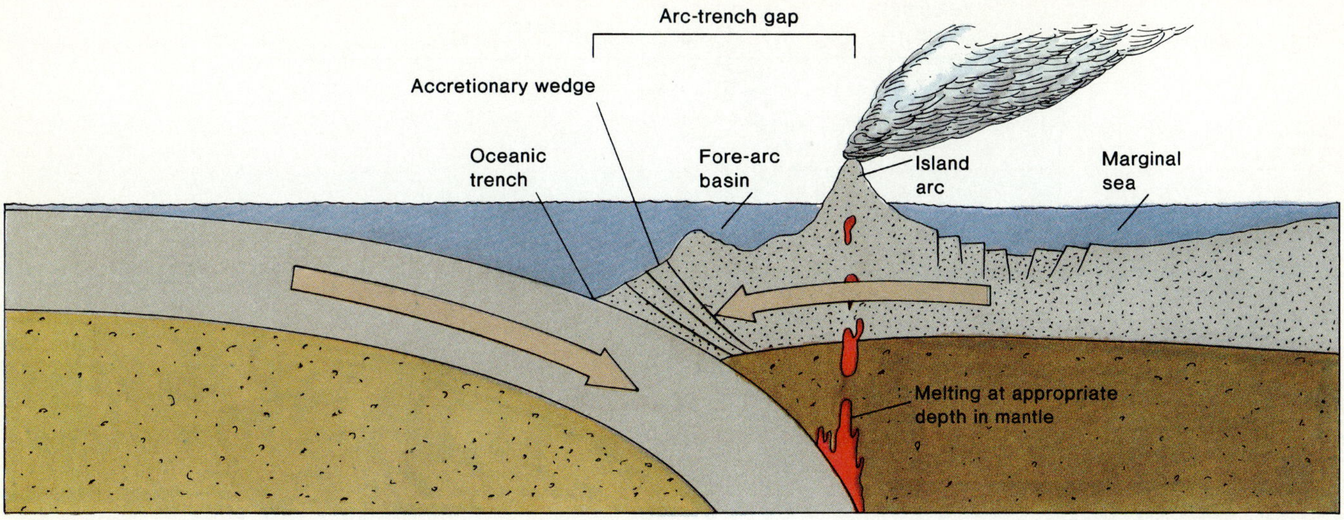

FIGURE 14.8 Profile across a subduction zone.

d) In figure 14.9a, sketch the upper and lower boundaries of the subducted plate, assuming that the deepest part of the Aleutian trench is at the middle of the bar representing the trench. Explain your reasoning.

e) At what depth does melting apparently begin beneath this portion of the arc? Explain your reasoning.

f) Assuming that melting occurs at the same depth beneath all segments of the arc, complete the other three profiles by drawing the subducted plate.

g) Measure the arc-trench gap in all four profiles.

Amchitka ______________ **Shumagin** ______________

Cook Inlet ______________ **Skwentna** ______________

h) Do the results agree with your predictions from question (a) on page 214? Explain.

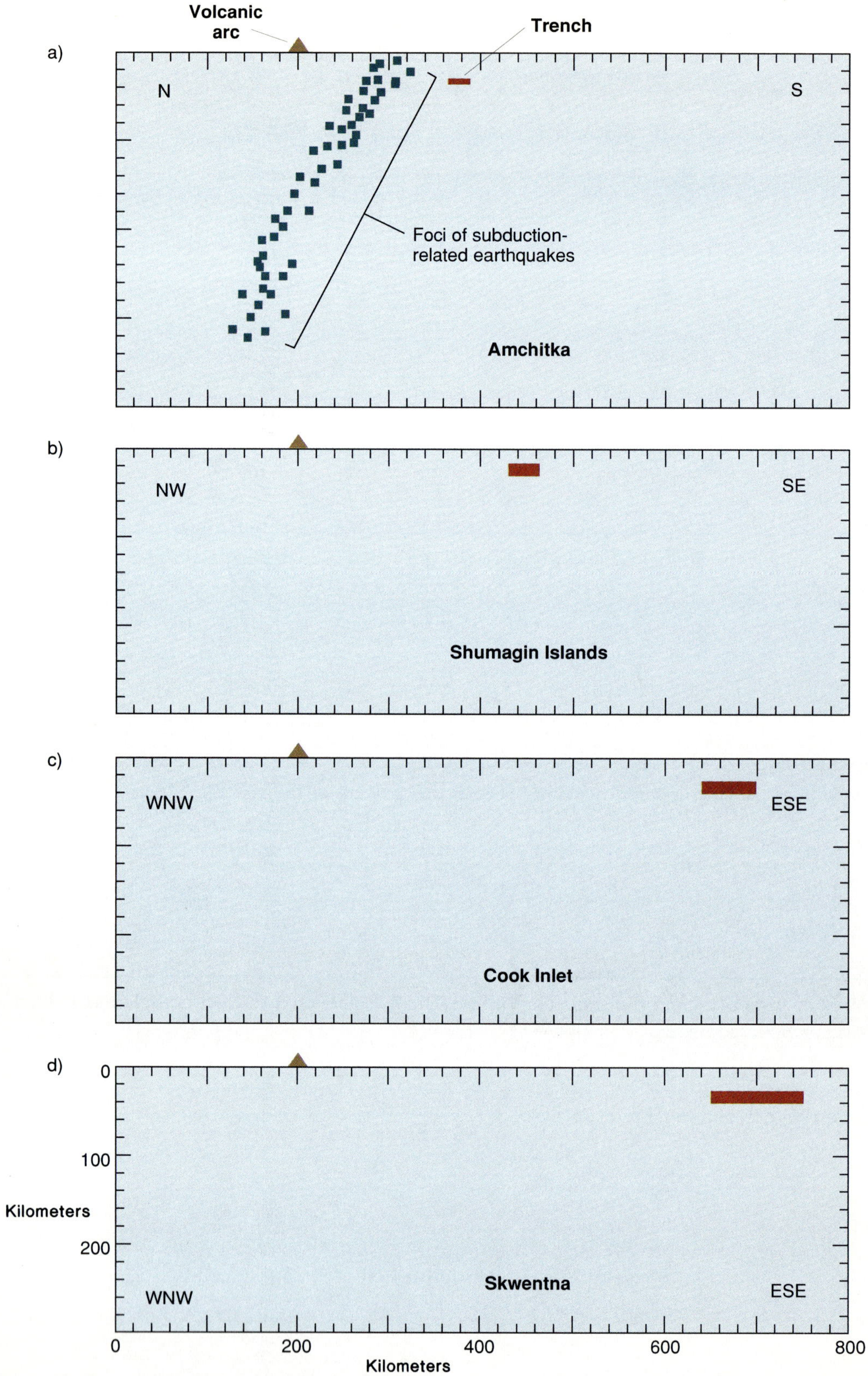

FIGURE 14.9 Profiles across segments of the Aleutian arcs.

Source: Jacob Nakamura and Davies, "1977 Trench-Volcano Gap Along the Alaska-Aleutian Arc: Facts and Speculations on the Role of Terrigenous Sediments for Subduction." in *American Geophysical Union, Maurice Ewing Series #1*, L. Talwani and Pitman, editors.

Re-examine figure 14.6, the physiographic map of part of the Pacific Ocean floor.

i) With a black pencil, outline as many ocean trenches as you can find. With a red pencil, outline the crests of the volcanic arcs associated with each.

j) Based on your answers to previous questions in this exercise, compare and contrast the nature of subduction in these subduction zones.

Re-examine figure 14.3, the physiographic map of the South Atlantic Ocean.

k) Use your black and red pencils as in (a) above to identify trenches and volcanic arcs.

l) The Andean volcanoes along the west coast of South America are also subduction related, but they are not a volcanic island arc. Explain.

m) Describe the differences between the tectonic activities currently going on in the South Atlantic and Pacific oceans.

CHAPTER 15

PUTTING IT ALL TOGETHER

PURPOSE

To use what you have learned from Sections 1, 2, and 3 to answer questions about regional and local geologic history specific to your campus and home.

MATERIALS NEEDED

1) A geologic map of North America or the United States.
2) A map showing the extent of Pleistocene glaciation in North America.
3) A geologic map of your state or province.
4) A geologic time chart showing relative and absolute ages.
5) Colored pencils.
6) Tracing paper.

INTRODUCTION

In chapter 13, you saw how it is possible to reconstruct aspects of the geologic evolution of large portions of North America from their general outcrop patterns. Different parts of the continent have had different geologic histories: some outcrops reveal billions of years of Earth history, but others only record a few tens of millions of years. Some areas have been intensely deformed in the collision between ancient continental plates, whereas others have remained parts of a single, relatively stable plate for hundreds of millions of years. Some areas have a "simple" history, while others are more complicated. The exercises in the chapter, combined with your own experience as a beginning field geologist in the area around your campus, are designed to help you learn a bit more about the geologic history and importance of your home and/or the place in which you are learning geologic principles. We will start with the large picture and then focus on your particular surroundings.

Exercise 15.1 Deciphering the Tectonic Evolution of North America

Figure 15.1 is an outline map of North America that shows political (i.e., national, state, province boundaries) and major streams. Use it in conjunction with the Geologic Map of North America to answer the following questions.

a) Where are the oldest rocks in North America? How old are they?

Every continent contains a broad area of ancient rocks, called a Precambrian Shield, that forms a continental core. Such areas tend to be tectonically stable today, and are often flanked by younger rocks that separate the shields from less stable, more tectonically active regions.

b) Sketch in the Precambrian Shield of North America on figure 15.1, and color it.

c) Based on the large-scale outcrop patterns shown on the Geologic Map of North America, identify the areas outside the Precambrian Shield where you would expect to find:

the most highly folded rocks

the least deformed rocks.

d) With which major topographic features are the most highly folded rocks associated?

e) With which major topographic features are the least folded rocks associated?

Folding, as we saw in chapter 12, is the result of intense compression, and regional-scale folding is generally the result of compression of a magnitude that can be produced only by plate collisions. Mountain systems formed by folding thus mark the zones of collision between ancient plates. Geologists say that plates have been "sutured" together in these zones.

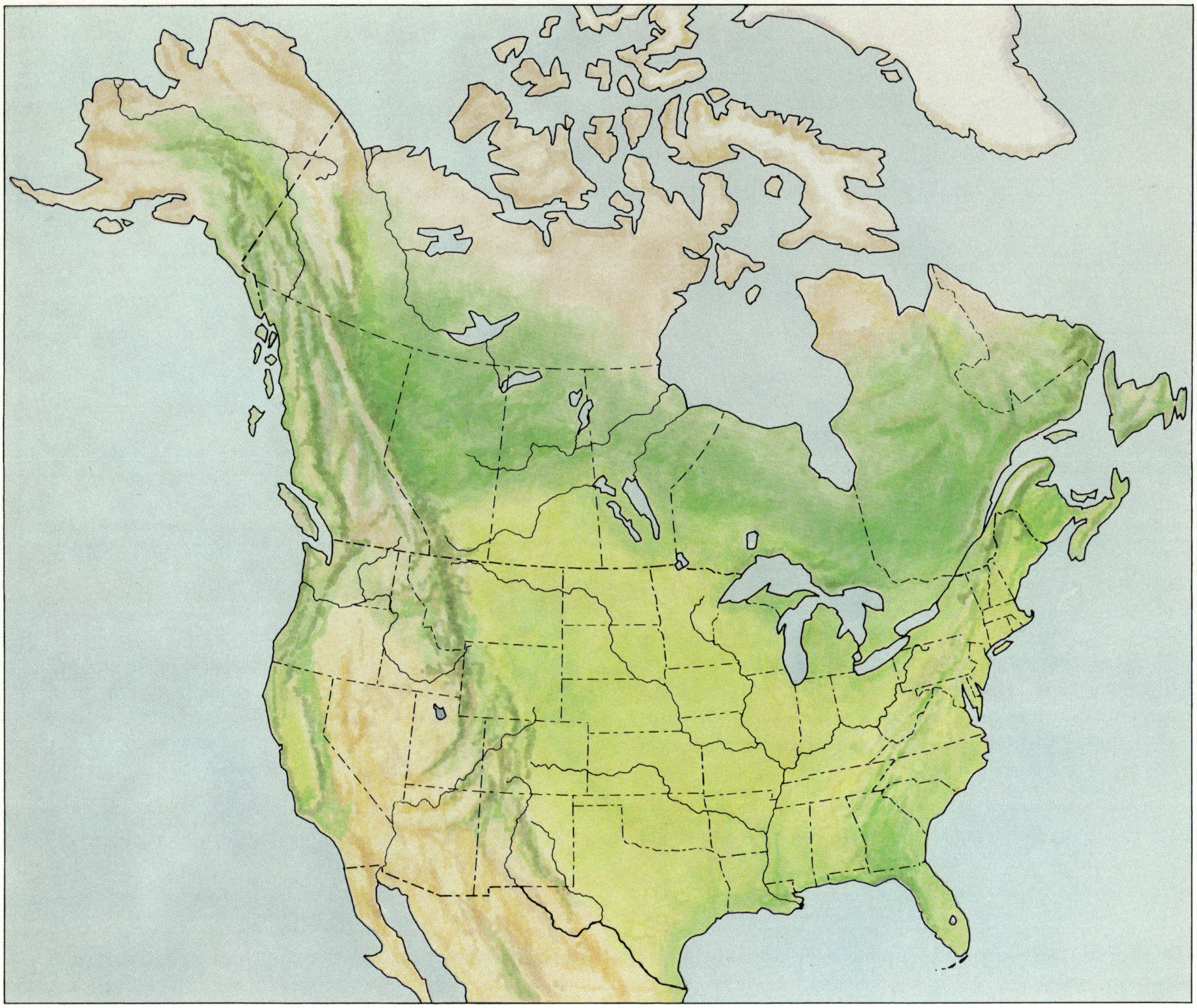

FIGURE 15.1 Outline map of North America.

---

f) Use a colored pencil to sketch these **suture zones,** using the central axis of the mountains as a first approximation of the zone.

Notice the position of the deformed zones around the Precambrian Shield. Areas separated from the Shield by these suture zones belonged to other continental masses in the geologic past, and were joined to ancestral North America by plate collisions over time.

It is now possible to determine when these different plates collided with the Precambrian core of ancient North America. You must simply figure out **when** each of the mountain systems formed. You have already done this in Exercise 13.3 for the Appalachian Mountains.

---

g) Carefully examine the rocks in the mountain systems that you outlined in f) above. How young is the youngest **deformed** rock in those mountains? How old is the nearest unconformably overlying undeformed rock? Fill in the table below:

TABLE 15.1 AGE OF MOUNTAIN-FORMING PROCESSES IN NORTH AMERICA

MOUNTAIN SYSTEM	AGE OF FOLDING

Note that not all mountain-building events took place at the same time. This suggests that plate collisions have occurred more than once, and that several separate plates have collided with the ancient core of North America. (NOTE: Of course, that core itself is the result of ancient collisions of even older plates, but those details are beyond our ability to resolve here.)

h) Based on the data in your table, color each plate that has collided with the Precambrian Shield with a different color.

i) Locate those areas where plate collision and transform faulting are going on today. Explain your evidence for these conclusions.

j) Assuming that present plate motions continue unchanged, what effect will these processes have on the size and shape of North America?

W A R N I N G

It is easy to fall into the trap of overgeneralizing, and to come up with the hypothesis that once a stable continental core has formed, a continent can only grow larger with time by 1) adding parts of other plates through subduction of oceans that once separated the plates, and eventual continent-continent collision; or 2) by transform-fault movement of plates from one place to another until they are sutured together. Remember that sea-floor spreading can split a continent apart, and that transform faulting can similarly split off a piece of a large continent, making it smaller.

EXERCISE 15.2 RELATIONS BETWEEN TECTONIC ACTIVITY AND EROSIONAL PROCESSES

In some instances, there is an easily recognizable relation between plate collisions, mountain building, and the subsequent type and nature of erosional processes. In others, the relations may be much more subtle. The following questions take a look at some of these relations.

Examine the major drainage patterns shown in figure 15.1 and on the Geologic Map of North America. Some rivers empty into the Atlantic Ocean either directly, as in the case of the Delaware River, or indirectly, through the Gulf of Mexico and Gulf of St. Lawrence. Others deliver their water to the Pacific Ocean (and Gulf of Baja California) or the Arctic (Hudson's Bay).

a) Place a piece of tracing paper over figure 15.1. Locate and sketch in the stream divides that separate the major drainage basins whose waters enter: a) the Gulf of Mexico; b) the Atlantic Ocean; c) the Pacific Ocean; d) the Arctic Ocean.

b) With which topographic features are these divides associated?

c) To what extent do the boundaries of ancient plates, and/or suture zones between them, control the large-scale stream erosional systems of North America?

d) Some streams do not drain into any ocean. Explain why, and locate places on the map of North America where you would expect to find such streams.

e) Locate the principal desert regions of North America on the tracing paper. To what extent are tectonic processes responsible for the locations of these regions of arid erosional activity? Be as specific as you can in your answer.

Examine the map showing the extent of continental glaciation during the Pleistocene Epoch.

f) Sketch the southern limit of Pleistocene glaciers on the tracing paper.

To what extent is this limit controlled by the boundaries or suture zones between ancient plates?

To what extent does the southern limit of glaciation control the major drainage systems?

g) Contrast and compare the degrees to which ancient tectonic activity controls stream and glacial processes in North America today and as far back as the Pleistocene. Suggest an explanation for any major differences.

Although continental glaciers covered a large portion of North America in the Pleistocene, much of the continent was not directly affected by the ice. Nevertheless, Pleistocene glaciation affected all of North America (and the rest of the world as well) in several ways.

h) Describe the ways in which Pleistocene glaciers affected climate and erosion in parts of North America that were not actually covered by ice.

i) The height of mountains can be used, to a first approximation, as an indication of their age. The principle is straightforward: old mountains have been attacked by erosional agents for many more years than young mountains. Using an atlas, large-scale relief maps of North America, or any other source of information concerning the elevations of mountains, which mountain systems are the oldest, and which are the youngest?

j) How do your conclusions compare with the ages of deformation and mountain-building that you determined in Exercise 15.1?

It would be tempting to try to establish the rate at which mountains are eroded by comparing the difference between the elevations of the old mountains and those of the young ones.

Explain what information you would need to do so that is not given in any of the sources for elevation.

What geologic problems are there that prevent you from establishing rates by such a method?

These connections and relations are only samples of those that can be interpreted from the rock and landscape records. Ties between such disparate areas as plate collisions and base level, stream patterns and subduction, and many others, can be made once you have the basic understanding of geologic processes provided by your course in Physical Geology. The more you know, the more connections seem to exist.

EXERCISE 15.3 GEOLOGY CLOSER TO HOME

The ability to answer the questions raised earlier reflects the degree of geological sophistication that you have attained in your Physical Geology course. As intellectually satisfying as your new knowledge and skills may be, it is perhaps even more rewarding to recognize how your newly acquired knowledge applies to the areas in which you live, grew up, and are going to school. The following questions are designed to show how your location fits in with the larger tectonic and geomorphologic pictures examined earlier in this chapter, and to show how your knowledge can be put to practical use.

Examine the geologic map of your state/province provided by your instructor. In some instances this will be a single standard sheet of paper; in others, a much larger, wall-sized map. The first set of questions deals with general features of the map and a comparison with the map of North America.

---

a) Compare the scale of your map with that of the Geologic Map of North America.

Does your state/provincial map provide greater detail or less detail than the North American map, based on scale alone? Explain.

Does the legend (explanation) for map colors and symbols provide more or less detail than the map of North America? Explain.

---

b) Give the latitude and longitude for the northernmost and southernmost parts of your state or province and for your campus.

---

c) Locate your campus as accurately as possible on figure 6.4b. If you held a compass outside your geology laboratory building, would the needle point east or west of geographic ("true") north? By approximately how many degrees?

The next questions refer to the relation of your state/province to the larger features of North America:

---

d) Are you located in the ancient core of North America or on one of the plates that was accreted to it after the Precambrian Era?

If the latter, when did your plate become part of North America? If the former, how far from the nearest ancient plate boundary are you located?

If your state/province contains a suture zone, on which side of the zone is your campus located? [HINT: Remember that young rocks may cover parts of the Precambrian Shield; locate your campus relative to major suture zones.]

---

e) In which major drainage basin (or basins) discussed in Exercise 15.2–1 is your state/province located? What is the master stream that delivers the water to the ocean? If your state/province contains a major drainage divide, in which basin is your campus located?

f) Is your campus located within or outside the area affected by Pleistocene continental glaciation?

If within, what effects of glacial action are visible in the area around your campus? If outside the glacial limit, how would you expect local climate and erosion to have differed in the past from what you now experience? What evidence of such differences have you observed on your local field trips?

g) Describe the age range of the rocks present in your state/province.

Assuming that the Earth is 4.6 billion years old, what percentage of Earth history is revealed by the rocks exposed at the surface in your state/province?

How far and in what direction would you have to travel to visit the nearest outcrops of rocks of Precambrian age? To find rocks deposited in the Cretaceous Period, when *Tyrannosaurus rex* roamed the Earth?

The last set of questions refers specifically to information about your state or province:

h) Determine the relative ages of rocks exposed in your state/province.

Where are the oldest rocks? What kind of rocks are they?

Where are the youngest rocks? What kind of rocks are they?

i) Based on your state/provincial map, its legend, and what you have seen on field trips, what do the types of rock and their ages tell you about the evolution of the area around your campus?

What was the source of sedimentary rocks in the region (or sources if more than one is present)?

What was the paleogeography at the time those rocks were deposited (shallow marine? deep marine? continental)?

j) Describe the deformational history of your state/province in as much detail as you can, based on the overall outcrop pattern of strata and the ages provided by the map legend. Supply evidence for each of your conclusions. Be sure to include such information as: deformed vs. undeformed; age(s) of deformation; type of deformation (folding, faulting); and nature and age of unconformities.

k) What kinds of geologic structure are most common in the area around your campus? Show how they relate to the larger scale deformation in your state/province.

l) With which episodes of plate collision (if any) was the deformation associated? (Use the ages determined in Exercise 15.1 to identify the events.)

m) Which agent(s) of erosion are responsible for the topographic features found on and around your campus?

To what extent does differential erosion of weaker and more resistant rock types contribute to the surrounding topography?

Which rock or sediment type(s) lie beneath the higher elevations? The lower elevations?

Would your answer to (b) be different if you lived in a different climatic zone? Explain why, on general geological principles, it might **be different, and then answer for the specific case.**

n) Describe how the geologic setting of your campus would affect attempts to build a new 12-floor building for the Geology Department. Include such features as the nature of the foundation for the building (bedrock vs. unlithified sediment); and geologic structures, topographic features, and specific rock or sediment types that might pose a problem in construction.

Assuming that price is no object (by no means a valid assumption!), and the outside of your new building is to be covered with a resistant, aesthetically pleasing material, what natural resources in the region could be used for facing stone?

Do the rock and/or sediment types present in the area make it easy to produce the materials needed to build a modern building? Remember, you will need most or all of the following:
Sand, gravel, and cement for concrete.
Gypsum for "sheet rock."
Clay for brick.
Aluminum for the latest in wall studs.
Copper for wiring.
Quartz-rich sands for making glass.

o) Once the building is finished and your Geology Department has moved in, will it get its water from unconfined aquifers in the area, confined aquifers, or reservoirs fed by local and regional stream systems? Explain in detail.

Credits

Chapter 1

Figures 1.1a, 1.1b, 1.1c, 1.1d, 1.1e, 1.1f, 1.2a, 1.3, 1.5: © Wm. C. Brown Communications, Inc./photography by Omni Resources, Inc.

Chapter 3

Figures 3.1a, 3.1b, 3.1c, 3.1d, 3.1e, 3.1f, 3.1g: © Wm. C. Brown Communications, Inc./photography by Omni Resources, Inc.

Chapter 4

Figures 4.2a, 4.2b, 4.2d, 4.2e, 4.2f, 4.3d, 4.3e, 4.3f, 4.3g, 4.3h, 4.4f: © Wm. C. Brown Communications, Inc./photography by Omni Resources, Inc.; Figure 4.2c: Courtesy of Wards Natural Science Establishment, Inc.; Figure 4.3a&b: © Wm. C. Brown Communications, Inc./photography by Omni Resources, Inc. (Thin Section)—© Wm. C. Brown Communications/Michael Peres, Photographer; Figure 4.3c: © Wm. C. Brown Communications, Inc./photography by Omni Resources, Inc. (Thin Section)—Geoscience Resources, Burlington, N.C.; Figures 4.4a.1, 4.4c.2, 4.4d, 4.4e: Geoscience Resources, Burlington, N.C.

Chapter 5

Figures 5.2a, 5.2b, 5.2c, 5.2d, 5.2e, 5.3a, 5.3b, 5.3c, 5.3d, 5.3e, 5.3f: © Wm. C. Brown Communications, Inc./ photography by Omni Resources, Inc.

Chapter 6

Figures 6.6a, 6.6b, 6.6c: U.S. Geological Survey

Chapter 7

Figures 7.3, 7.4, 7.5, 7.7, 7.8: U.S. Geological Survey

Chapter 8

Figures 8.3, 8.8, 8.9, 8.10: U.S. Geological Survey; Figure 8.5: New York State Department of Transportation; Figure 8.6: © Mitchell Algus; Figure 8.7: Dyke & Prest, 1987, Geological Survey of Canada Map, 17024

Chapter 9

Figures 9.3, 9.4: U.S. Geological Survey

Chapter 10

Figures 10.4, 10.5, 10.8, 10.9: U.S. Geological Survey

Chapter 11

Figures 11.6, 11.7, 11.8, 11.9, 11.10, 11.11, 11.12, 11.13: U.S. Geological Survey

Chapter 12

Figures 12.3a&b, 12.9: U.S. Geological Survey; Figure 12.10: Virginia Division of Mineral Resources Report of Investigations, No. 2; Figure 12.11: Geologic Map

Chapter 13

Figures 13.7, 13.8, 13.9, 13.10, 13.11: Geologic Map; Figure 13.12: NASA

Chapter 14

Figures 14.3, 14.6: BRUCE C. HEEZEN AND MARIE THARP, 1977, © MARIE THARP, 1977 1 WASHINGTON AVE, SOUTH NYACK, NY. 10960

Illustrator

BioGraphics

Figures 1.4, 1.6, 1.7, 1.8, 2.1, 3.2, 3.3, 4.1, 5.1, 6.1, 6.2, 6.3, 6.4a&b, 6.5, 6.7, 6.8, 6.9, 6.10, 6.11, 6.12, 6.13, 6.14, 7.2, 7.6, 7.9, 8.2, 8.4, 9.1, 9.5, 9.6, 9.7, 9.8, 10.3, 10.6, 10.7, 11.3, 12.1, 12.2, 12.4, 12.5, 12.6, 12.7, 12.8, 13.3, 13.4, 13.5, 13.6, 14.2, 14.4, 14.5, 14.7, 14.8, 14.9, 15.1